EFFECTIVE MATHEMATICS LESSONS THROUGH AN ECLECTIC SINGAPORE APPROACH

Yearbook 2015

Association of Mathematics Educators

EFFECTIVE MATHEMATICS LESSONS THROUGH AN ECLECTIC SINGAPORE APPROACH

Yearbook 2015

Association of Mathematics Educators

Wong Khoon Yoong

Published by

World Scientific Publishing Co. Pte. Ltd.
5 Toh Tuck Link, Singapore 596224
USA office: 27 Warren Street, Suite 401-402, Hackensack, NJ 07601
UK office: 57 Shelton Street, Covent Garden, London WC2H 9HE

British Library Cataloguing-in-Publication Data
A catalogue record for this book is available from the British Library.

EFFECTIVE MATHEMATICS LESSONS THROUGH AN ECLECTIC SINGAPORE APPROACH
Yearbook 2015, Association of Mathematics Educators

ISBN 978-981-4696-41-8

Printed in Singapore

In memory of Glen T. Evans,
eminent Australian teacher educator and my PhD supervisor.
May his legacy continue through the educators he had taught and inspired.

For Zoe, Emma, Isabelle, and their future mathematics teachers

Foreword

Singapore's stellar performance in the Third International Mathematics and Science Study of 1995 and continued good performance in Trends in International Mathematics and Science Studies of 2003, 2007 and 2011 has brought a lot of attention to the teaching and learning of mathematics in Singapore schools. Numerous reports and papers written about the why, what and how of this has yet to provide a one stop resource that may enlighten anyone seeking answers to questions pertaining to the classroom pedagogy of mathematics teachers in Singapore schools.

Dr Wong Khoon Yoong, an eminent mathematics educator in Singapore, was invited by the Association of Mathematics Educators to write the 2015 Yearbook of the Association. Dr Wong accepted the invitation and has written this book that provides readers with a much valued insight of teaching and learning of mathematics in Singapore schools. It comprises a range of learning experiences and teaching strategies that mathematics teachers can judiciously select and adapt in order to deliver effective lessons to their students at the primary to secondary level. The Singapore mathematics framework, which has stood the test of time for more than two decades and is internationally recognized, forms the bedrock of the book.

We, the colleagues of Dr Wong and members of the Association of Mathematics Educators in Singapore, are truly indebted to Dr Wong for his invaluable intellectual contribution to the mathematics education community in Singapore and the world.

Berinderjeet Kaur
National Institute of Education
Singapore

Acknowledgements

This book is the result of several favourable conditions. I wish to thank the following people who have brought about these conditions.

The Executive Committee (2014 – 2016) of The Association of Mathematics Educators, Singapore, for inviting me to prepare this yearbook of the Association under a new format. Without this invitation, this book would not have been written.

Berinderjeet Kaur for coordinating this project, writing the Foreword, and reviewing the manuscript. Toh Tin Lam and Toh Pee Choon for raising helpful points about mathematical and teaching ideas that may be misunderstood due to ambiguous statements.

Low-Ee Huei Wuan for meticulously correcting many typographical errors. Ultimately I am solely responsible for any shortcomings found in this book.

Martin Greenacre, my son-in-law, for giving many insightful comments from an engineer's perspective, alerting me to several recent education reports, and discussing the influences of education reforms on mathematics learning.

Wong Yee Jun, my son, for reading part of the manuscript, also from an engineer's lens.

Peter Galbraith, my PhD co-supervisor, and many scholars whose work has enriched my intellectual life and classroom praxis.

Teachers who have attended my classes for teaching me that the written words need to be experienced before they can be successfully applied to classroom teaching and learning.

Last but not least, Teh King Choo, my wife, for her strong support especially during the past few months of hectic writing and for sharing our love and life together.

Contents

Chapter 1

Curriculum: Map the Intended, Implemented, and Attained Landscape

Curriculum defines the goals and learning experiences organised in a formal way by education systems and schools to educate their students in various subjects. There are three types of curriculum: the intended, the implemented, and the attained. This chapter examines these three types of curriculum for mathematics and links them to the nature of mathematics as a discipline.

Mathematics is the surest way to immortality. If you make a big discovery in mathematics, you will be remembered after everyone else will be forgotten. Paul Erdös (1913 – 1996)

1 Nature of Mathematics

Every school subject has its unique disciplinary structure which must be taken into account when one designs its school curriculum. Briefly, mathematics is frequently associated with numbers and logical thinking, science with phenomena and experimentation, languages with communication and imaginative writing, history with chronological records of events and people, and so forth. By studying these subjects, students gain discipline-based knowledge and distinctive styles of thinking so that they can make sense of the world from different perspectives and more importantly, respond coherently to real-life situations by integrating what they have learned from the relevant disciplines. In terms of Gardner's Theory of Multiple Intelligences (2004), it is not enough to develop just the seven or nine types of intelligence separately; rather, the challenge is to be able to effectively combine them to tackle specific problems.

The fundamental question, then, is, "What is mathematics?" In their classic book with this title, Courant and Robbins (1947) gave the following answer:

> Mathematics as an expression of the human mind reflects the active will, the contemplative reason, and the desire for aesthetic perfection. Its basic elements are logic and intuition, analysis and construction, generality and individuality. (p. xv)

Their answer highlights a balance of dichotomies: active vs. contemplative, logic vs. intuition, analysis vs. construction, generalisation vs. specialisation, pure vs. applied, and beautiful vs. ugly mathematics. Thus, there are different aspects of the nature of mathematics. Philosophers of mathematics describe these aspects as foundations of mathematics using different types of *–isms*:

- *Platonism*: mathematics exists in an ideal universe, which Erdös called the Book [1], and mathematicians discover some of its contents.
- *Formalism*: mathematics is a game of rules, just like chess, made up by mathematicians. This is most evident in the construction of axiomatic systems such as different types of geometry. This aspect stresses that mathematics is created rather than discovered.
- *Logicism*: mathematics can be reduced to logic. The most famous attempt was by Whitehead and Russell to prove the truth of all mathematics using axioms and symbolic logic in their *Principia Mathematica* project. However, Gödel showed that these attempts have failed (see Mackenzie, 2012). Nevertheless, people often associate mathematics with logical thinking.
- *Empiricism*: mathematicians discover mathematical properties through analysing facts gathered from sensory experiences, similar to scientific experimentation; this conforms to the etymological root of mathematics as science or craft.

---

[1] http://en.wikiquote.org/wiki/Paul_Erd%C5%91s

- *Intuitionism*: mathematics must consist of objects that can be constructed; in this case, the idea of infinity would be problematic.
- *Ethno-mathematics* or *multi-cultural mathematics*: mathematics is a social construction, and mathematicians from different cultural groups, for examples, Arabic, Chinese, Indian, and European, develop different criteria for truth claims and different types of mathematical thinking. This evolves from the work of D'Ambrosio (2006). According to Ernest (1991), mathematics is "a living social construction, with its own values, institutions, and relationship with society in the large" (p. 107).

Although mathematics teachers need not know these –isms in depth, they must be aware of these ideas because this knowledge helps them to understand their own beliefs about mathematics, as well as those held by their students. Bishop (1991) delineated six types of activities that are mathematical in nature: counting, measuring, locating, designing, explaining, and playing. These different ideas about the nature of mathematics can be arranged into a taxonomy, shown in Table 1.1. They are placed at different grade levels to reflect the mathematics maturity of the students when they advance in their mathematics journey. In this taxonomy, mathematical aspects at earlier levels must still be continually deepened at higher levels.

Teachers can better understand these aspects of the nature of mathematics if they try to identify the nature of common mathematics tasks. An exercise for this is shown in Figure 1.1. Teachers who did this exercise could classify most of the items according to the categories in Table 1.1. Most of them thought that item 3 is about equal sharing, but they did not have other ideas about fairness. They disagreed that item 8 is mathematical, and the problems they posed for item 9 were mostly rule-based.

Their answers to exercises similar to the above reveal their beliefs about mathematics. These beliefs affect their thinking about teaching mathematics, in particular, their answers to important curriculum questions such as the ones below:

- Why is it important to teach mathematics in schools?
- What mathematics is important to teach?

- Is a particular activity mathematical or just busy work?
- How to teach mathematics effectively?
- What is important to assess in mathematics learning and how to assess it in mathematically meaningful way?

Table 1.1
Nature of mathematics related to grade level

Grade Levels	Nature of Mathematics	Remarks
Primary	Set of rules	Use numbers collected by counting, measuring, or estimating; apply rules to real-life situations; compute with or without tools
	Hierarchy of concepts	Logical sequence and connections across concepts
	Inductive thinking	Patterns and relationships, intuition
Secondary	Deductive thinking	Proof, logical reasoning
	Language	Communicate and explain ideas using symbols and specialised terms
	Applications beyond mathematics, problem solving	Modelling physical and social worlds; learning of other subjects across the curriculum; see the world through mathematical lenses
	Visualisation	Mental imagery, graphics, arts
	Games	Puzzles, enjoyment
Post-secondary	Axiomatic system	Implications of different axioms
	Creativity	Discover or create new concepts, rules; problem posing, conjecture
	Mathematical attributes	Rational, open-mindedness, precision, perseverance, democratic participation
	Ethno-mathematics	Mathematics in cultural artefacts; mystical properties of mathematical objects, e.g., Platonic solids
	Programming	Algorithmic thinking, link to computer science

1. Solve $2x^2 - 3x - 5 = 0$.
2. Veterinarians estimate that kittens grow to adulthood in one year. Each year after that, one "cat year" is equivalent to six "human years". Assume that human adulthood starts at 18 years. What is the relationship between "cat year" and "human year"?
3. Divide \$90 fairly among 3 persons. In how many ways can you do this?
4. Prove that if $xy = 0$, then either $x = 0$ or $y = 0$, where x and y are real numbers.
5. Is the probability that someone has cancer given that the test is positive the same as the probability that the test is positive if someone has cancer?
6. Babylonian method: To find $\sqrt{15}$, (a) make a guess; (b) divide 15 by the estimate; (c) find the mean of answers to (a) and (b); (d) repeat step (b) and so on.
7. A restaurant offers a discount based on the age of the diner. For example, a 40-year-old diner will be given 40% discount on the bill. What is the discount if the diner is 104 years old?
8. Why is the 50 cent coin larger than the one dollar coin in Australia?
9. Make up a problem about this pattern and solve it.

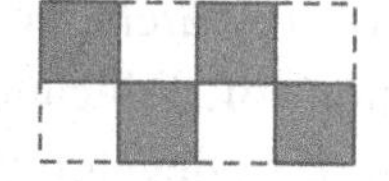

Figure 1.1. Identify the mathematical nature of tasks

2 Three Types of Curriculum

The term *curriculum* (plural: *curricula* or *curriculums*) comes from the Latin word *currere*, meaning a *racing track* or *course*. UNESCO (2013)[2] supplied a succinct definition: "a description of what, why and how students should learn." The *what* and *why* are also called the *intended* curriculum, and the *how* is the *implemented* curriculum. In addition to these two types of curriculum, educators include a third one called the *attained* (or *achieved*) curriculum. This deals with *how well* or *how much* the students have learned from the implemented curriculum.

---

[2] UNESCO International Bureau of Education (IBE), http://www.ibe.unesco.org/fileadmin/user_upload/Publications/drafts/IBE_curriculum_glossary_final.pdf

The race track metaphor is an apt one to depict these three types of curriculum:

- *Intended* curriculum; this is a blueprint of the track to define what students are expected to learn according to official specifications. These specifications are organised by grade levels, age groups, key stages, or bands; they correspond to the different types of racing tracks.
- *Implemented* curriculum; this is the actual track. It describes what the students actually experience in the classrooms and around the school.
- *Attained* curriculum; this covers what the students actually know and can do as a result of the teaching they have received. This is similar to the outcomes of the race, in particular the winners and losers.

The curriculum framework comprising these curriculum types is used in many research projects. The International Association for the Evaluation of Educational Achievement (IEA) has been using this framework to develop the research questions for its international studies in mathematics and science education (e.g., Robitaille & Garden, 1996). School teachers can use this framework to unpack the mathematics curriculum offered in their schools. To do so, they need to identify the curriculum elements that make up each of these curriculum types. The essential elements are summarised in Table 1.2.

US educators tend to use the term *curriculum* to mean textbooks and teaching materials and the term *standards* for what the students should understand and be able to do, that is the topics. For example, the US Common Core for State Standards in Mathematics (CCSSI, 2010) placed mathematics topics under the *Standard* label. The term *standards-based curriculum* is used to identify textbooks and resources that are aligned with the topics and competencies advocated by US educators and institutions, such as the NCTM *Standards* (2000).

Table 1.2

Types and elements of different curriculums

Curriculum Types	Main Themes	Elements of a Subject Curriculum
Intended	Why	• Goals, aims, objectives of the curriculum, at different levels: national, regional, schools, classes • Framework, philosophy underpinning the curriculum • Alignment with national goals of education, e.g., 21st Century Competencies • Alignment with school mission, e.g., to nurture pupils to be gracious citizens and creative thinkers
	What	• Topics, scopes, scheme of work • Cross curriculum contents
Implemented	How	• Sequence of topics, pacing, lesson plans • Teaching resources, textbooks, technology • Grouping of students into streams, bands • Pedagogy
Attained	How well How much	• Formative assessment • Summative assessment • Ipsative assessment

Romagnano (2006) clarified the various meanings of *standard* as follows:

- "content standards are the mathematical strands that compose the school mathematics curriculum" (p. 7); this is the *what* of the UNESCO definition;
- "performance standards are the achievement expectations to which students are held" (p. 8); this is the everyday meaning about quality or level of achievement; it is related to the attained curriculum;
- "benchmarks are descriptions or models of specific performance standards" (p. 9); this is also about the attained curriculum.

To this list one might include *implementation* standards to cover the quality and scope of opportunities to learn (OTL) offered to the students. These OTL are often assessed in research and curriculum evaluation as indicators of the implemented curriculum. To avoid confusion, educators should specify which meanings of curriculum and standards they are writing about. This book will use the meanings captured in Table 1.2.

Textbooks provide a link between the intended and implemented curriculum, and their role in curriculum studies is rather ambiguous. Textbooks and teaching resources serve as an important part of the intended curriculum at school level when the teachers refer to them for the scope of the topics to be covered. On the other hand, when teachers strictly follow the explanations, worked examples, and exercises given in the curriculum materials, they become part of the implemented curriculum. To highlight this conceptual conundrum, the authors of the TIMSS analysis of mathematics textbooks referred to them as "potentially implemented curriculum" (Valverde et al., 2002, p.13).

Curriculum documents in many countries also recommend specific pedagogies, which is part of the intended curriculum. Again when teachers actually implement them, even with some modifications, these strategies become the implemented curriculum.

The above curriculum elements are formal, explicit, and organised, but informal types of curriculum are also of interest. A *hidden* (or *implicit*) curriculum refers to what are learned without explicit instruction, perhaps through peer influences, and these often become student beliefs about schooling or learning. Teachers need to be aware of these learnings, which tend to be more negative than positive. For mathematics learning, the hidden curriculum includes the following rather negative perceptions:

- Always look for *the* right answer to every problem.
- Workings must be neat.
- Do the problems quickly.
- Use only the methods taught to solve different types of problems. Do not try to mix the methods.
- Mathematics is mysterious and students have no control over how the rules work.

Null curriculum refers to topics not included in the intended curriculum. For example, matrix transformation becomes a null curriculum nowadays in the Singapore curriculum, although it was a favourite topic in the past. The rush to cover the syllabus is called the "hurried curriculum" (Hargreaves et al., 2001), and a *learner-designed curriculum* is one designed by the students themselves (Hyde, 1992), which is not a common practice.

Curriculum changes with time, as evident from the numerous curriculum reforms undertaken by many countries in the past decades. There are two trends in these reforms, and these are vividly captured in the classic fable called the *saber-tooth curriculum* (Benjamin, 1939/2004). In this fable about Palaeolithic education, the educational goal was to ensure that children have better food, shelter, clothing, and security in the future. This goal was to be achieved by training them to handle real-life, authentic tasks in three areas: to scare saber-tooth tigers with fire, to club woolly horses, and to grab fish with bare hands. In each case, the children were put in the actual situations so that they learned as they lived. Centuries later when the Ice Age arrived, the saber-tooth tigers became extinct, the woolly horses ran away, and the river became muddy. Now, the community was invaded by bears and antelopes. These environmental changes led to a tension between two camps of educators. Its *wise* educators wanted to keep the old curriculum because they believed that it could teach children generalised skills such as agility through using fire, clubbing horses (even though no real ones were available for practice now), and fish grabbing. The other group of educators saw the need to revise the curriculum to teach new skills, such as to dig bear pits, to use snares to catch antelopes, and to catch fish with nets rather than bare hands. Contemporary curriculum reformers face similar conflicts. In mathematics education, a current conflict is whether to teach paper-and-pencil operations such as fraction addition, when these can now be easily carried out using calculators. No common decision has been reached within and across education systems.

In addition to the official curriculum, there is the *tuition curriculum*, a major component in the education system of most East Asian countries. In Singapore, this is a billion-dollar industry and serves the needs of students of different abilities. The high achieving students take tuition to prepare them to get into special programmes; the average students want to keep ahead of school lessons; the low achieving students require additional help just to keep up with the lessons and to improve their grades. Thus, tuition is another element of the hidden curriculum: school teaching is inadequate, especially for mathematics as it is probably the most popular tuition subject.

The next section will look at the formal mathematics curriculum, beginning with the *why* question.

3 Intended Mathematics Curriculum: *Why*?

Mathematics is a core subject in most education systems around the world. Several reasons are advanced to support this world-wide practice. The most basic reason (goal or aim) for school mathematics is to develop students' ability to effectively use mathematics knowledge and skills in daily life. The Singapore mathematics curriculum (Ministry of Education, 2012) expressed this goal in this way: "acquire mathematical concepts and skills for everyday use and continuous learning in mathematics" (p. 10). The goal of continuous learning is differentiated for vocational education, sciences, and other subjects under different mathematics programmes at higher grade levels. This approach is apt because it underscores the point that the same goal can be realised differently at different grade levels.

In recent years, many new terms have been used to cover this goal with nuanced differences: *functional numeracy, quantitative literacy, mathematical literacy,* and *mathematical competencies.* However, educators do not always agree on what these terms mean. Among these terms, mathematical literacy has gained wide international acceptance because it is used in the PISA (Program for International Student Assessment) studies. It is defined for PISA 2012 (OECD, 2013) as follows:

> Mathematical literacy is an individual's capacity to formulate, employ, and interpret mathematics in a variety of contexts. It includes reasoning mathematically and using mathematical concepts, procedures, facts and tools to describe, explain and predict

phenomena. It assists individuals to recognise the role that mathematics plays in the world and to make the well-founded judgments and decisions needed by constructive, engaged and reflective citizens. (p. 25)

This OECD definition recognises that the utility goal has values for the individuals as well as the society. Chapter 5 will elaborate on this goal with examples of mathematical applications.

A comprehensive list of curriculum goals is given in Table 1.3. Notice that "passing public examinations" is included at the bottom of the table. Educators tend to consider this as an assessment goal rather than a curriculum goal. But for education systems where public examination is the critical component, teaching to the test has become the major curriculum goal as well, and this fact has to be recognised in curriculum design.

It is possible to rate every goal in Table 1.3 based on its importance and the current opportunity to learn it at different grade levels. This rating can be based on practical experiences, research, and theories about the stages at which mathematical cognition develops. For example, van Hieles theory (1986) suggests that deductive proof should be given low rating at primary level because it is beyond most primary school students. On the other hand, daily application has high rating at primary levels because many basic mathematics skills taught at these levels can be applied directly to real-life situations, whereas more abstract topics at higher grade levels have limited daily applications.

A 3-point rating scale (1 = *low*, 2 = *medium*, 3 = *high*) is used in Table 1.3. The values are suggestive rather than definitive. Reading across each row, one considers the rating of the same goal at different grade levels, as mentioned above. Down each column, the ratings suggest relative emphasis given to different goals for the same grade level. Of course, these ratings reflect the philosophy of mathematics education held by the raters. It is a worthwhile exercise for teachers in the same school to compare their ratings to arrive at common understanding about how their teaching can help to achieve these broader curriculum goals. Education systems around the world will emphasise these goals differently, and a comparative study along this direction can contribute to international understanding.

Table 1.3

Goals for mathematics curriculum across grade levels

Curriculum Goals	Primary	Secondary	Post-Sec
Daily applications of skills and knowledge, utilitarian and functional	3	1	1
Further learning, across subjects, preparation for courses and careers	3	3	3
Mathematical problem solving: unfamiliar, novel, extended, modelling	1	2	3
Make informed decisions based on mathematics	1	1	2
Generalisation from patterns, inductive thinking	3	2	2
Logical and coherent reasoning, proof	1	2	3
Understand mathematical language	2	2	2
Enjoyment, confidence, interest	3	2	2
Metacognition	1	2	2
Alignment with national goals, e.g., 21CC	1	2	2
Pass public examinations	3	3	3

The goals in Table 1.3 are broad and long-term ones. They cannot be accomplished within a few lessons or by isolated topics. Teachers need to plan lessons so that they can achieve these goals. They should remind their students of these long-term goals as the latter move up the grade levels. Such reminders can provide strong motivation to students over many years and may even arrest the deterioration in student attitudes towards learning mathematics with grade levels, which has been found to be almost a universal trend. See Chapter 7 for further discussion about attitudes.

4 Intended Mathematics Curriculum: *What*?

The *what* aspect of the intended mathematics curriculum covers two dimensions: mathematics contents and mathematical processes. These two dimensions are given different names and coverage; see Table 1.4 for those used in Singapore, TIMSS (Mullis, 2012), and CCSSI (2010).

Table 1.4

Content and process dimension of intended mathematics curriculum

Dimensions	Singapore	TIMSS	CCSSI
Content	• Number and Algebra • Measurement and Geometry • Statistics	• Grade 4: Number, Geometric shapes and Measures, Data display • Grade 8: Number, Algebra, Geometry, Data and Chance	• Standards (sub-topics); Clusters; Domains • High schools: Number and Quantity; Algebra; Functions; Modelling; Geometry; Statistics and Probability
Processes	• Reasoning, Communication and Connections • Applications • Thinking Skills and Heuristics	• Knowing • Applying • Reasoning	• 7 mathematical processes; see Table 1.5 below

This diversity of terminology and classification of the *what* of different intended curricula does not allow for easy comparison across countries and research projects. One way that may facilitate such comparison is to organise details about topics and processes under a curriculum framework.

5 Intended Mathematics Curriculum: *Curriculum Framework*

A curriculum framework serves the important function of linking the different curriculum elements together to help the users better grasp its major and essential focus. However, not many mathematics curricula around the world have such frameworks in their documentation. Indeed, in a recent review of the Australian national curriculum, the reviewers (Donnelly & Wiltshire, 2014) noted that the lack of an "overarching curriculum framework" is "the missing step" in the development of the subject contents (p. 2). The development of the Singapore mathematics curriculum may shed light on the process of designing and maintaining such a framework over many years (Lee, 2008; Wong & Lee, 2009).

The curriculum review committee[3], tasked to design the Singapore mathematics curriculum in late 1980s, realised the importance of having an overarching framework to convey to the teachers the gist of mathematics education for Singapore students. A coherent framework also mitigates the common perception that mathematics is merely a collection of isolated facts, concepts, and rules for different topics. The framework it produced is sometimes called the *Pentagon Framework*. It is probably the longest surviving curriculum framework in any school subject throughout the world.

The framework places problem solving as the central goal of mathematics education. This was based on the official requirement at the time to make the ability to solve problems a key education outcome for Singapore students. Two major publications in the 1980s also supported this goal and were consulted. They were the Cockcroft Report (1982) and the NCTM yearbook (1980). The Cockcroft Report stated that "the ability to solve problems is at the heart of mathematics" (p. 249) and the NCTM made problem solving its topmost recommendation for school mathematics in the 1980s[4]. However, the committee decided to define *problem* in an inclusive sense to include routine exercises, problems in unfamiliar contexts, open-ended investigations, and real-life problems.

---

[3] Members: Chang See Tong (Chair), Goh Mei Lian, Sin Kwai Meng, Kho Tek Hong, Lim Chee Lin, Dolly Chong, Wong Khoon Yoong

[4] http://www.nctm.org/standards/content.aspx?id=17278

This focus on problem solving has not changed for the past twenty five years, and it is conceivable that it will remain so for many years to come.

The committee also saw the need to summarise the many factors that can promote problem solving into a few constructs to help teachers form a holistic perspective of and navigate through the curriculum. After much deliberation, it decided on five components or factors: *concepts, skills, processes, attitudes,* and *metacognition*. Each of these five inter-related components is an important learning outcome on its own, and together they help students become better problem solvers. Over the years, minor changes were made to some of the components, but the overall structure remains intact. Figure 1.2 shows the latest version (Ministry of Education, 2012). An important lesson from this curriculum reform is that an education team should think very carefully about the major thrust of the reform it intends to bring about and should allow considerable time for the stakeholders to find ways to achieve the intended goals. Frequent changes to the goals and the curriculum structure are likely to be counterproductive.

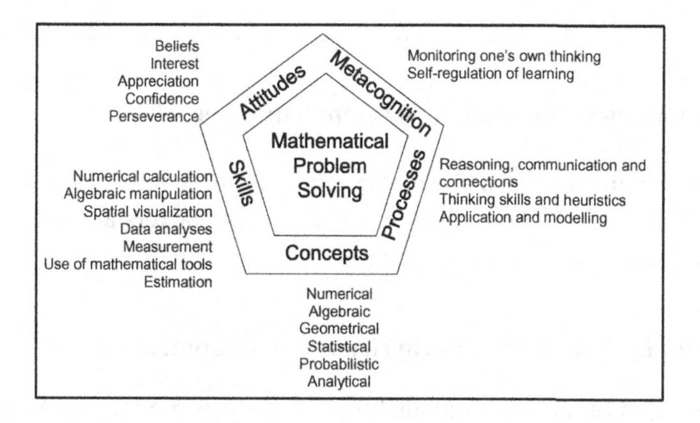

Figure 1.2. Singapore mathematics curriculum framework

There are striking parallels between the Pentagon framework and two US frameworks. This is shown in Table 1.5. One notable observation of the entries is that the labels of the problem solving factors used in the Singapore framework are simpler than those in NRC (2001). This simplicity allows teachers to discuss mathematics learning with other stake-holders, especially parents of their students.

Table 1.5
Comparison of mathematics curriculum frameworks of Singapore and US

Singapore (1990)	NRC (2001)	CCSSI (2010)
Skills	Procedural fluency	• Attend to precision
Processes	Adaptive reasoning	• Reason abstractly and quantitatively • Construct viable arguments and critique the reasoning of others • Look for and express regularity in repeated reasoning
Metacognition	Strategic competence	• Make sense of problems and perseveres in solving them • Use appropriate tools strategically • Model with maths
Concepts	Conceptual understanding	
Attitudes	Productive disposition	

6 Mathematics Curriculum Development: *How to*?

The development of a mathematics curriculum is influenced by many interlocking factors. Two ways to examine these factors are discussed in this section.

6.1 Strands of mathematics curriculum development

This is based on the classical analysis of Howson, Keitel, and Kilpatrick (1981); see also Howson (1983). They classified mathematics curriculum development into five strands (*movements, forces* or *approaches*) based on different emphasis on mathematics contents and psychological theories. The predominating features of these strands are summarised below.

a) *New math.* This was promoted by mathematicians who focussed on unifying concepts, structures, and axioms that form the foundation of mathematics. Students learn to state properties using precise mathematical language, such as the Commutative Law and the Associative Law. In the 1980s, Syllabus D taught in Singapore schools had elements of this strand.

b) *Behaviourist.* The common practice of breaking mathematical concepts into hierarchies is a hallmark of programmes based on this strand. It includes several laws of behaviourism enunciated by Thorndike, Skinner, Gagné, and others. Despite negative comments about behaviourism by some educators, its influence on mathematics instruction remains potent. Direct instruction, computer-based instruction, and individualised instruction are just some examples of its enduring influence.

c) *Structuralist.* This strand takes note of the structure of the disciplines and is based on the work of Bruner and Dienes, in particular. Programmes under this strand stress discovery learning, modes of representation as embodiments of mathematics, principles of variability, and the spiral curriculum. The CPA (Concrete → Pictorial → Abstract) approach used in Singapore mathematics instruction is based on Bruner's three modes called the enactive, iconic, and symbolic modes. Singapore has also designed its curriculum based on the spiral approach. One of Bruner's well-known comments about the goal of teaching mathematics is:

> We teach a subject not to produce little living libraries on the subject, but, rather, to get a student to think mathematically for himself [sic], to consider matters as a historian does, to embody the process of knowledge-getting. (1964b, p. 335)

d) *Formative.* This term refers to development or growth in cognition, and the key force behind this strand is Piagetian Theory of Cognitive Development. It explains how mathematical cognition and attitudes develop at different ages through interactions with the environment. Special features include use of concrete manipulatives at the concrete operational stage and attention to *readiness* in intellectual development.

e) *Integrated-environmentalist.* This multi-disciplinary movement includes real-world problem solving and modelling. Many recent mathematics reforms include elements of this strand.

These strands are not static. According to Galbraith (1988), they seem to occur in cyclic patterns over decades in the following way: applications, pure mathematics, psychology, technology, structure, back to basic, and mathematics. Another way to describe the changes in curriculum reforms was by Lambdin and Walcott (2007) in this way: drill and practice, meaningful arithmetic, new mathematics, back to basics, problem solving, standards, and assessment and accountability. A powerful emerging curriculum is ICT-enabled, including computer programming. It arises from the ways in which ICT has recently changed mathematical practices. This strand will be explored again in Chapter 6.

6.2 Situated Socio-Cultural Framework

The development of a curriculum in general and in mathematics is affected by local contexts and global influences. The interplay between these two spheres of influence is articulated in a *situated socio-cultural framework* proposed by Wong, Zaitun, and Veloo (2001). This framework was used to explain why Brunei Darussalam, Malaysia, and Singapore have taken different paths in developing their mathematics education in the past five decades, even though they had similar mathematics curriculum during their colonial past under the British system. This framework may also explain the development of mathematics curriculum in other countries, an idea yet to be tested through detailed case analysis. A concise version of the framework is shown in Figure 1.3.

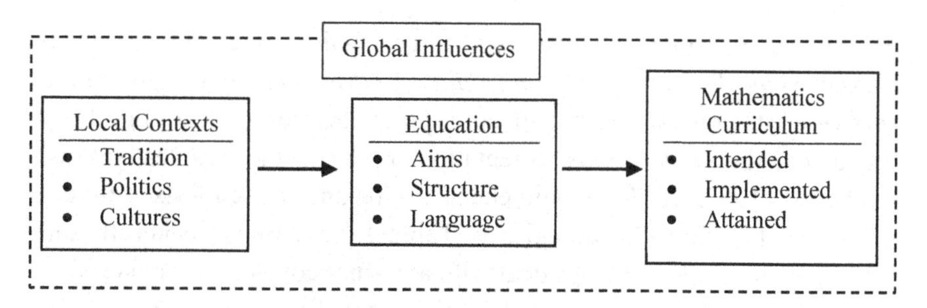

Figure 1.3. Situated Socio-cultural Framework

The locally situated factors include relevant traditions, politics, and cultures. They have direct influences on how the national education system works, especially in terms of national aims of education, its structure, and the language of instruction. The mathematics curriculum, being part of the national system, is determined by these prior forces. Most countries are nowadays affected by the highly globalised environment, and this forms the broad umbrella under which the local, situated factors will function. Curriculum developers need to be able to derive locally relevant best practices by critical analysis of global trends. The key elements within each set of factors have differential strengths, and the challenge that curriculum developers face is how to take advantages of these strengths to design a mathematics curriculum best suited to the needs of the students at a particular point in time.

7 Implemented Mathematics Curriculum: *How*?

Pedagogy is the art and science of teaching and it is made up of many different teaching or instructional strategies. The implemented curriculum refers to the actual use of these strategies in the lessons.

Research has identified the effects of different teaching strategies on learning from the West and East perspectives. In a widely-cited second order meta-analysis of research across different school subjects, Hattie (2009) reported that teacher variables contributed about 30% of variations in student achievement, and the most impactful strategy is to give meaningful feedback to the students. From mathematics education,

Askew et al. (1997) identified three major orientations called *transmission, discovery,* and *connectionist,* corresponding roughly to the behaviourist, constructivist, and conceptual theories of learning. They found that the *connectionist* orientation produced the best test results. Cockcroft (1982, #243) recommended the balanced use of six different strategies. The first six strategies in Table 1.6 are from Cockcroft, and the remaining three strategies deal with areas not covered by Cockcroft.

These strategies can be used to teach the five components in the Singapore framework. Table 1.6 also shows the chapters in which these strategies are to be discussed in detail. This organisation is a heuristic for writing, and it does not mean that a given strategy can be used for the particular components mentioned in the table. For example, teacher exposition is necessary to teach all the five components, but it is covered in detail in Chapters 2 and 3.

Table 1.6

Main strategies for teaching mathematics and their coverage in different chapters

No.	Strategies	Chapters and Components
1.	Exposition by the teacher in a whole-class setting	2: Concepts; modes of representation, graphic organisers 3: Skills; direct instruction
2.	Discussion between teacher and pupils and between pupils themselves	4: Processes: Discussion and enhanced Question and Answer
3.	Appropriate practical work	2: Concepts
4.	Consolidation and practice of fundamental skills and routines	3: Skills; mastery, information processing
5.	Problem solving, including the application of maths to everyday situations	4: Heuristics 5: Applications and modelling
6.	Investigational work	4: Processes
7.	Reading mathematics texts and writing about mathematics	8: Metacognition
8.	Internet search	6: ICT
9.	Motivating students	7: Attitudes, motivation

These teaching strategies need to be combined with the specific mathematical topic at a fine-grained level to become the maths-related pedagogy. Consider the explanation strategy. Giving examples is an essential feature of clear explanation in general, but the examples for fraction addition are very different from those for decimals, even though they both belong to the same Number topic. This consideration applies to the other strategies in Table 1.6.

To underscore this distinction, the construct *pedagogical content knowledge* (PCK) introduced by Shulman (1986) is helpful. It covers

> the most useful forms of representation of those ideas, the most powerful analogies, illustrations, examples, explanations, and demonstrations-in a word, the ways of representing and formulating the subject that make it comprehensible to others. ... includes an understanding of what makes the learning of specific topics easy or difficult: the conceptions and preconceptions that students of different ages and backgrounds bring with them to the learning of those most frequently taught topics and lessons. (p. 9)

For mathematics, this becomes *mathematics pedagogical content knowledge* (MPCK). It is a complex intersection of mathematics content knowledge, pedagogical knowledge, knowledge of students, and knowledge of the mathematics curriculum. It is believed that teacher's MPCK will affect the actual implemented curriculum for their students. Several studies have produced tests to assess MPCK; some examples are the work by Felmer et al. (2014); Hill, Ball, and Schilling (2008); Lim-Teo et al. (2007); and Tatto et al. (2012).

As an illustration of MPCK, the framework used in the Teacher Education and Development Study in Mathematics (TEDS-M) is discussed here (Tatto et al., 2012). This international study involved 17 countries and about 22,000 future teachers. The study assessed the mathematics content knowledge and MPCK of these future teachers. The research framework divides MPCK into the three sub-domains shown below. The first sub-domain is about the intended curriculum, and each of the remaining two sub-domains combines implemented and attained curriculum. Sample skills for each sub-domain are also included:

- *Mathematics curriculum knowledge*, e.g., know the school mathematics curriculum, establish appropriate learning goals, identify key ideas in learning programs.
- *Knowledge of planning for mathematics teaching and learning* (pre-active), e.g., select appropriate activities, identify different approaches for solving mathematical problems, choose assessment formats and items.
- *Enacting mathematics for teaching and learning* (interactive), e.g., generate fruitful questions, diagnose student responses including misconceptions, and provide appropriate feedback.

One of the released TEDS-M MPCK items for primary school future teachers is shown in Figure 1.4. It was an Enacting item.

[Jeremy] notices that when he enters 0.2×6 into a calculator his answer is smaller than 6, and when he enters $6 \div 0.2$ he gets a number greater than 6. He is puzzled by this, and asks his teacher for a new calculator! (a) What is [Jeremy's] most likely misconception? (b) Draw a visual representation that the teacher could use to model 0.2×6 to help [Jeremy] understand WHY the answer is what it is?

Figure 1.4. An TEDS-M enactment item

For (a), about 26% of the international sample and 54% of the Singapore sample could provide the two ideas underlying this misconception. The respective results for (b) were 23% and 39% (Wong, Boey, et al., 2012). Teacher educators can use similar findings to plan appropriate activities for their methodology courses to help future teachers strengthen their MPCK.

8 Attained Curriculum: *How Well*?

The attained curriculum refers to the learning outcomes, which cover cognition, affect, psychomotor skills, and social behaviours. This requires the collection of evidence about these outcomes using different types of assessment tasks and processes. One important insight from educational assessment is that testing in whatever form can measure only a portion of what is intended and actually implemented in the lessons. Hence, the attained curriculum is a subset of the other two types of curriculum.

Education assessment covers a vast area of research, praxis, and complex and confusing terminology (e.g., see the glossary from University of Tasmania[5]). This section covers only basic ideas.

8.1 *Assessment goals*

Assessment is used for a variety of goals. The three main goals are:

- Assessment *of* learning (AoL) or summative assessment; this *measures* student performance at the end of a time period. The most obvious form is public examinations or standardised tests.
- Assessment *for* learning (AfL) or formative assessment; information about student's work is used to plan future learning and teaching activities. Many AfL situations should be used in order to capture a variety of curriculum goals. Diagnostic assessment, as a form of AfL, is a popular way used to identify student errors and misconceptions in specific mathematics topics.
- Assessment *as* learning (AaL); this involves students in the learning-cum-assessment process, so that the teacher and student together set personalised learning goals, clarify success criteria, and use self- and peer-assessment to collect evidence about outcomes. See Earl (2013) and Mooney et al. (2012).

8.2 *Quality of assessment*

Mathematics assessment can take many forms, including quizzes, tests, projects, practical work, oral presentations, and homework. Quality assessment tasks and procedures should have many of the following characteristics:

---

[5] http://www.teaching-learning.utas.edu.au/assessment/terminology

- Correct mathematical contents and reasoning; some commercially produced and teacher-constructed tests have wrong mathematics.
- Alignment with the curriculum, covering an adequate representative sample of the goals and contents.
- Alignment with instruction in terms of opportunity to learn the contents.
- Validity; the test measures what it is designed to measure (traditional meaning) or the test data are interpreted and used according to the intended purposes (AERA, APA, & NCME *Standards,* 1999).
- Reliability; the data are collected under conditions that minimise measurement errors to ensure consistency of students' responses.
- Fairness, lack of bias against sub-groups of students in terms of gender, socio-economic backgrounds, language use, familiarity of the tools used in the assessment (e.g., hand-writing, calculator use, computer entry); however, students' differing ability per se is *not* a concern about fairness because the test is set to measure this ability.
- Clarity; precise mathematical language is used and words used to describe real-life contexts are within the expected literacy competence of the students.
- Balance of easy and difficult items, preferably arranged in graded order of difficulty with the easy ones at the beginning and the difficult ones at the end.
- Facility index and discrimination index are within standard ranges.
- Usability, practicality, and efficiency of test administration, including security for high-stakes examinations.

8.3 *Interpretations of assessment data*

The same set of assessment data can be interpreted in three different ways: criterion-referenced, norm-referenced, and ipsative-referenced. Sometimes the same data are interpreted in all the three ways; this is not recommended. Figure 1.5 gives a pictorial representation of these interpretations.

Criterion-referenced testing (CRT or standards-based)
The test is based on a set of pre-determined criteria that align with the curriculum aims and contents. The test data are interpreted to show how much of the assessment domain has been mastered by each student, without referring to the performance of other students in the same cohort. Most school tests are criterion-referenced using marking schemes that define the criteria at different standards.

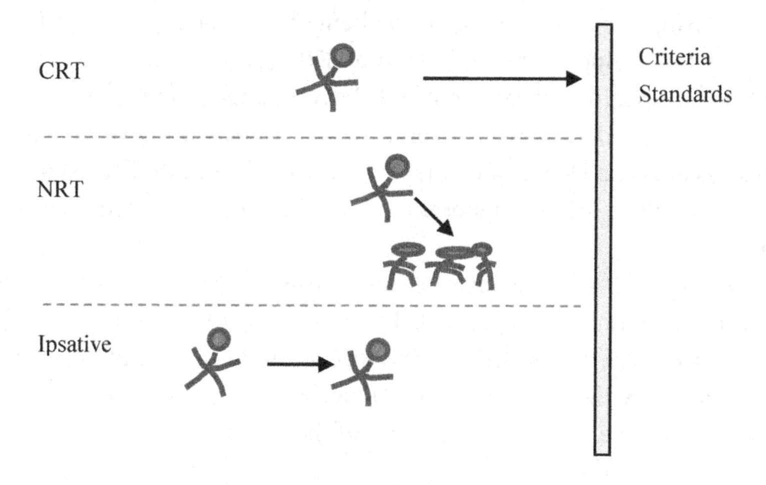

Figure 1.5. Three interpretations of assessment data

Norm-referenced testing (NRT or normative assessment)
A student's achievement is interpreted in comparison to that of other students in the same or similar cohort. Raw scores are scaled to percentile scores, stanine scores, or T-scores in order to rank the students. Norms such as *mathematics age* are based on the Normal distribution, and cut-off scores are used to define different grades or attainment bands. Most public, high-stakes examinations are NRT.

Ipsative-referenced testing (or student-referenced)
The performance of a student in the test is interpreted in relation to his/her previous performance. This attempts to measure learning progression. Hogan (2007) used the term *self-referencing repeated*

measures. It is difficult to make this inference for academic tests including mathematics because students rarely repeat the same cognitive test, unlike the case with physical skills such as running a 100 metre race, where the student can practise the skill repeatedly and see improvement over time. A portfolio of student work in lieu of a single test may be used to make inferences about progress. For example, in England, under the *APP* (Assessing Pupil's Progress) framework, primary school teachers make overall level judgement about pupil's progress using observations and mathematical work (DCSF, 2010). Under the US Response to Intervention (RtI) reform, a student who is placed under specific intervention will have his/her progress monitored using short tests administered several times a week over several weeks. The results are then checked against an *aim-line* for the student (Bakken, 2012). Esquith (2013) supported this kind of interpretation and recommended:

> Make sure your students measure their progress against their own past performance, not other people's. Teach them that there will always be a faster runner, better writer, or more accurate calculator of math. Show them by their previous work that they are getting better, and do not hold up other kids' work in front of them. (p. 159)

In addition to the quantitative data used to make the above three types of interpretation, teachers should collect qualitative data by analysing students' errors, interviewing them about their mathematical and problem solving processes, observing their interactions with peers, using checklists to probe beliefs and perceptions, and so on. Students may also rate themselves or their peers on success criteria. These qualitative data provide useful data, to complement quantitative scores, to assess the quality of the attained curriculum.

Although the attained curriculum typically refers to students, the same idea can be applied to teachers. Data about teacher practices can be collected to ascertain to what extent the intended curriculum has been attained. In Hattie's (2005) words,

> If students do not know something, or cannot process the information, this should be cues for teacher action, particularly teaching in a different way (the first time did not work!). Merely ascribing to the

student the information that they can or cannot do something is not as powerful as ascribing to the teacher what they have or have not taught well. (p. 17)

9 Concluding Remarks

This broad survey covers the three types of mathematics curriculum and emphasises that teachers should not lose sight of the broad aims of the curriculum in the midst of preparing and teaching individual lessons. The different aspects of the nature of mathematics are related to curriculum goals and maths-based pedagogy or MPCK.

The Singapore mathematics curriculum is based on a succinct and stable framework that highlights the five major components required to help students become better problem solvers. These five components will be covered separately in Chapters 2 to 8.

The first component is about *concepts,* because they are the essential ingredients of mathematics. The Singapore framework places this component at the base of the Pentagon, stressing its foundational role in the teaching and learning of mathematics. Students need to make sense of mathematics concepts expressed in multiple modes of representations and to see the connections among concepts. Teachers have to find ways to help their students achieve this conceptual understanding. Several ways are discussed in the next chapter.

Chapter 2

Concepts: Build Meanings and Connections

Mathematics concepts are the foundations on which are built the mathematics skills, proofs, reasoning, and applications. This chapter delineates the essential features of concepts in terms of meanings, examples, non-examples, modes of representation, and conceptual connections. Strategies for concept formation and concept mastery using multiple modes of processing are discussed.

Mathematics is the tool specially suited for dealing with abstract concepts of any kind and there is no limit to its power in this field.
Paul Dirac (1902 – 1984)

1 Hierarchies of Concepts

Mathematics comprises numerous hierarchies of concepts and their inter-relationships expressed as theorems, formulae, principles, and algorithms. Some common hierarchies in school mathematics are shown below (some concepts are left out in these hierarchies):

- Numbers → even and odd numbers → prime numbers → integers → rational numbers → … → real numbers → complex numbers
- Variables (unknowns) → expressions → equations → solutions
- Shapes → squares → quadrilaterals … → circles
- Area → length × breadth → … → curved surface area
- Weight (or mass) → kilogram → density
- Chance → equally likely outcomes → probability of single events → probability of compound events → … → probability distributions

The Singapore mathematics curriculum delineates six types of concepts, and these concepts can be matched to the corresponding skills and content strands. This is shown in Table 2.1. Since the Singapore curriculum does not explain what are *analytical concepts,* I have included these under Calculus, which is an optional topic for the more able students in upper secondary levels. The hierarchies given above can be easily tagged with one of the six types of concepts.

Table 2.1
Concepts, skills, and content strands in the Singapore mathematics curriculum

Concepts	Skills	Content Strands
Numerical	Numerical calculation Estimation	Number and Algebra
Algebraic	Algebraic manipulation	
Geometrical	Spatial visualisation Measurement Use of mathematical tools	Geometry and Measurement
Statistical	Data analysis	Statistics and Probability
Probabilistic		
Analytical		Calculus

The teaching of concepts should cover three key aspects: meanings, modes of representation, and conceptual connections. These three aspects are inter-related, but for ease of discussion, they are dealt with in separate sections below.

2 Meanings, Examples, Non-examples

Every mathematical concept has a definitive meaning and this is illustrated with the use of examples and non-examples. Direct teaching of a concept has to cover these three aspects.

2.1 *Meanings*

The meaning of a concept is usually given as a definition in terms of *primary* or lower-order concepts in a hierarchy (Skemp, 1971). Every concept is associated with its own terminology, symbol or diagram. For example, one definition of a *prime number* is "a natural number with only two factors, 1 and itself." For this definition, the primary concepts are *natural number, factor,* and *1.* The meanings of these primary concepts must be understood in order to make sense of the given concept.

Students with poor knowledge of the relevant primary concepts will have difficulty making sense of a concept through definition. Hence, it is not recommended to give students at lower grade levels mathematically rigorous definitions when they are first introduced to a new concept. Instead, cognitive psychological theories recommend the use of practical experiences with concrete materials to help students make sense of (or even discover) the attributes of primary concepts. At this stage, imprecise wording is frequently used. Once students have built up sufficient mathematics vocabulary, they can learn the secondary or higher-order concepts through definitions. Even then teachers need to check that the students have acquired the correct meanings through definitions.

To illustrate, the following well-known activity is used to teach the concept of prime numbers. Begin with the numbers 2 and 6, as shown in Figure 2.1.

a) Count out the number of inter-locking blocks for each number.
b) Arrange the blocks to form a rectangle in as many ways as possible, treating rectangles with the same length and width as equivalent.
c) The length or width of each rectangle is a factor of the number. In Figure 2.1, the factors of 2 are 1 and 2, and of 6 are 1, 6, 2, and 3.
d) Classify each number according to the number of ways found in step (b). Those with only one possible arrangement are called prime numbers (e.g., 2), while those with more than one arrangement are called composite (or non-prime) numbers (e.g., 6).

e) Repeat the above process with other numbers. Classify them into primes or composites.

f) Encourage students to talk about the pattern using the new concepts and terms.

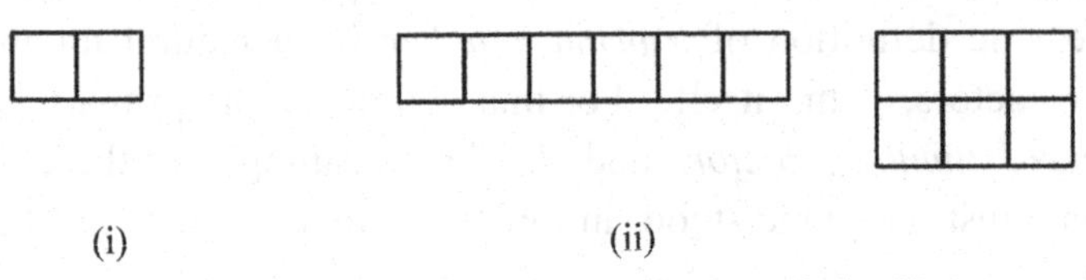

(i) (ii)

Figure 2.1. Concept of prime numbers: (i) 2 (ii) 6

A concept may be defined in slightly different ways, and students need to realise this practice and be able to check that definitions that are worded differently have the same meaning. A prime number may be defined as "a natural number which can be divided evenly by 1 and itself only." This definition does not use the *factor* concept; instead the concept of *divide evenly*.

Another example of this common practice is about the definition of an even number expressed in words or in symbols:

- A number which can be divided evenly by 2.
- A number of the form $2n$, where n is a natural number.
- A number with 2 as a factor.

When students first learn a new concept through activity, they should be encouraged to state the meaning in their own words. Their statements may be imprecise, and this is acceptable if the meaning is valid and can be understood by other students. The aim is to first develop the idea before focussing on its precise mathematical statements, which can come later.

Although mathematics is often presented as a precise subject, some key concepts, especially the primary ones, have different definitions in subtle ways. These differences appear in different textbooks. This discrepancy has caused considerable confusion among teachers and students. It is important to maintain the same meanings of and symbols for key concepts throughout a course until they are changed for certain purposes. Students must be alerted to different usage by textbook authors and mathematicians, so that they would pay greater attention to different definitions of the same concept given in other sources.

To illustrate this confusion, consider the definition of an isosceles triangle: a triangle with two equal sides. Here, *two* may mean *exactly two* (the *exclusive* sense used in daily life and by Euclid, see Artmann, 1999) or *at least two* (the *inclusive* sense). Given this ambiguity, whether the statement "An equilateral triangle is isosceles" is true or false depends on the definition used. The purpose of asking such statements of students is to assess their ability to make inferences based on given definition, but some teachers expect students to memorise these statements as if they were always true; see Section 4.2.

The idea that inferences are to be based on given definitions is a fundamental aspect of what Galbraith (1982) called *mathematical vitality.* He devised the following item to test this idea among pre-service teachers.

> Definition: A quadrilateral is the figure obtained by joining four points *A, B, C, D* in a plane by the straight lines *AB, BC, CD, DA*.
> Statement *S*: The angles of a quadrilateral at *A, B, C, D* sum to 360°.

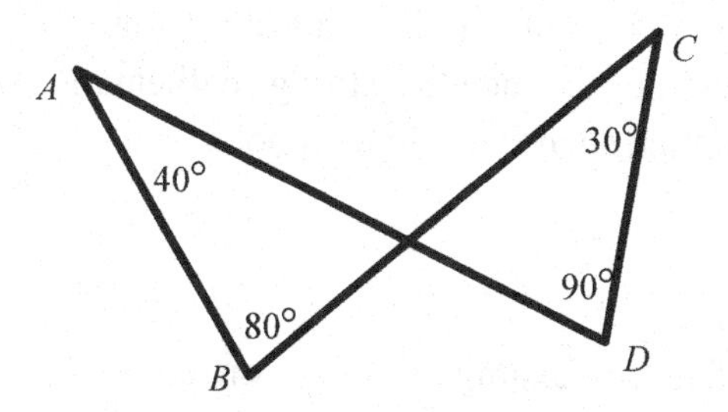

Figure 2.2. "Quadrilateral" (Galbraith, 1982)

About 71% of Australian (n = 116) and 76% of Singaporean (n = 74) pre-service teachers (Wong, 1990b) thought that the given figure is not a quadrilateral, because it conflicts with the usual image that the edges of a quadrilateral do not intersect. Only about 10% of each sample correctly noted that the given figure *is* a quadrilateral (as defined), which refutes statement *S*.

Two weaknesses are evident here and need to be addressed in mathematics teaching especially at the upper grade levels:

- the inability to make valid inferences based on given definitions;
- lack of acceptance of alternative definition that is different from what one already knows; mathematical definitions can be arbitrary.

Although mathematical definitions may be somewhat arbitrary, they are so defined to serve certain purposes. Students be inducted into this mathematical practice. For example, 1 is neither prime nor composite to ensure that the Fundamental Theorem of Arithmetic holds. Another example is that a^0 is defined to be 1 ($a \neq 0$) so that it is a consistent extension of the laws of indices for natural numbers. However, it is wrong to think that one can *prove* definitions. A particularly striking misconception among teachers is to try to prove that $\pi = \frac{\text{circumference}}{\text{diameter}}$. What need to be proved is that this ratio is a constant, and its value is denoted by the symbol π.

Finally, some textbook definitions are incomplete. For example, a textbook gives this incomplete definition: a quadratic expression is one in which the highest power of x is 2. It does not state that it must be a polynomial. Teachers need to acquire strong mathematics knowledge in order to be able to notice such flaws in textbooks.

2.2 *Examples*

Examples (also called *yes* examples) of a concept must satisfy all the defining attributes of the concept but vary in non-defining ones. A series of examples should be arranged to include variations in increasing complexity, according to Dienes' *Mathematical Variability Principle* and *Perceptual Variability Principle* (Dienes, 1964; Gningue, 2006). The levels of complexity can be changed by incorporating the following factors:

- Size of numbers; larger values are more complex.
- Types of numbers: natural numbers → simple, positive fractions or decimals → negative integers → surds → real numbers → complex numbers.
- Number of terms, more terms being more complex.
- Types of operations; addition and multiplication → subtraction and division → power; expansion is easier than factorisation; differentiation is easier than integration.
- Number of operations, with and without brackets.
- Orientation of geometric shapes in diagrams; many geometric shapes in textbooks and teacher-designed worksheets are drawn with the bases horizontal (may be called the *canonical* orientation) and shapes in *slanted* orientation are quite rare.
- Real-life contexts add complexity due to additional demands on language and real-world knowledge.

These variations can be used to build a coherent set of examples, which is more effective than examples given in an ad hoc manner.

Example 1: Quadratic function, $ax^2 + bx + c$
Defining attributes: $a \neq 0$
Non-defining attributes: b and/or c may be zero
The following is a coherent set of examples:

- $2x^2 + 3x + 5$ [small, positive values]
- $2x^2 - 3x + 5$ [small, negative values]
- $x^2 + 3x + 5$ [$a = 1$, not shown and not *visible* to some students, so it should be discussed]
- $3x^2 + 26x - 46.5$ [decimal, large, negative values]

Example 2: Prism
Defining attributes:

- Two congruent n-sided polygons called bases.
- The n [lateral] faces joining corresponding sides of bases are parallelograms.

Non-defining attributes:

- The bases need not be regular; if they are regular, it is called a *regular* prism; in Figure 2.3(a), the bases are shown upright rather than flat and this helps students see that *base* in mathematics does not always mean something lying on flat surface.
- The lateral faces need not be perpendicular to the bases; if they are perpendicular, it is called a *right* prism; see Figure 2.3(b).
- Given the above two definitions, ask students to explain a *right regular prism* and draw an example as one way to test their conceptual understanding.

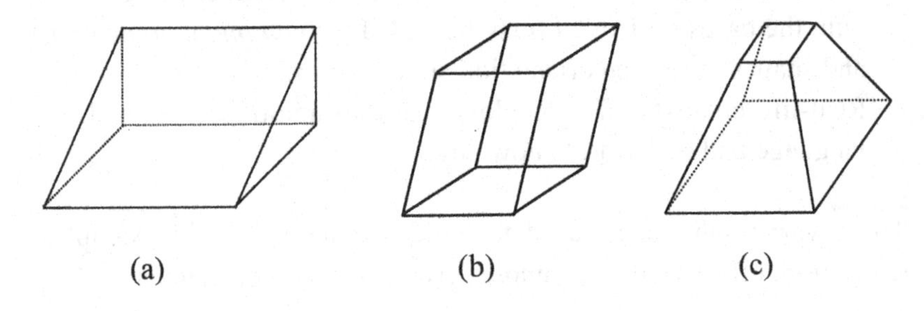

Figure 2.3. Examples of prism (a) & (b); Non-example of prism (c)

2.3 *Non-examples*

Non-examples (also called *counter-examples* or *no* examples) can help students make sharper distinction between defining and non-defining attributes of a concept. This engages students in the compare-contrast mode of thinking.

Suitable non-examples should share some but not all of the defining attributes of the concept. Such non-examples can be created by:

- contradicting some of the defining attributes; for prism, the two bases are not congruent, as shown in Figure 2.3(c);
- including additional attributes; for quadratic function, adding an xy term will make it not quadratic.

However, removing some features may or may not lead to non-examples. For quadratic function, removing the x^2 term results in a non-example, but removing the x term or the constant term will not make a non-example.

Suitable non-examples may be related to misconceptions about the concept. For example, if the set of numbers {3, 5, 13, 4, 20, 32} are used in activity about primes, students may form the misconception that prime numbers are always odd and composite numbers are always even. In this case, non-examples will help students learn by confronting their misconceptions. Providing only examples but not non-examples is ineffective.

2.4 *Frayer Model*

The definition/examples/non-examples components of a concept can be summarised using the Frayer Model (Krpan, 2013). The target concept is placed in the centre of a piece of paper, which is divided into four quadrants covering *Definition, Facts/Characteristics, Examples,* and *Non-examples.* Figure 2.4 shows the Frayer Model for *highest common factor* (HCF) (*Greatest Common Factor, GCF).*

The four examples in Figure 2.4 are chosen based on the following considerations.

- The first set involves an even and an odd number; this is quite typical.
- The second set involves two even numbers (could be two odd numbers) but the first number is larger; this counters the tendency to always give the smaller number followed by a larger one.
- The third set shows that HCF could be 1; this happens when the given numbers are *relatively prime,* a new concept which may be introduced here.
- The fourth set shows that HCF could equal the smaller of the pair of given numbers.
- Further examples should involve finding the HCF of three or more numbers.

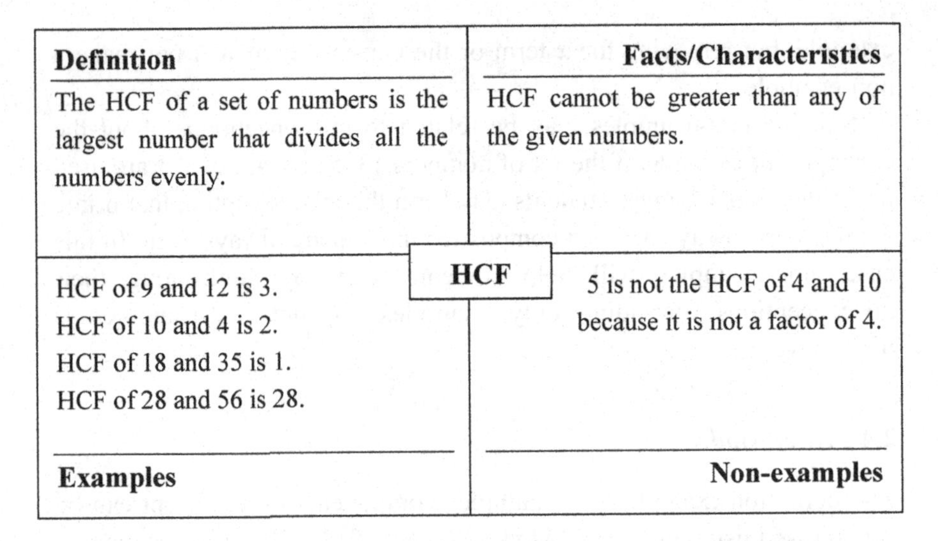

Figure 2.4. Frayer Model for highest common factor (HCF)

For active learning, students should generate their own examples and non-examples (e.g., Watson & Mason, 2005) and to complete the Frayer Model using it as an aid for memory and revision.

3 Modes of Representation

3.1 *Functions of representations*

Mathematical concepts are abstract and reside in the brain. The internal representations of mathematical ideas have to be externalised in various forms so that they can be operated on and communicated to other people. These constitute different *modes* of representation. According to Zazkis (2011), under different representations, certain information becomes more or less transparent. Table 2.2 shows how this idea works when the same number is written in different representations. These representations can convey different properties which can be used to answer different questions about the same number. Only samples of possible questions are given in the table. This is a worthwhile exercise to help students deduce properties from the ways numbers are written.

Table 2.2

Transparent vs. opaque properties of number modes

Number Modes	Transparent	Opaque
361	How many digits does it have in decimal form? Is it even or odd?	Is it prime or composite?
19^2	Is it a square number? What are its factors?	How many digits does it have in decimal form?
$3 \times 7 \times 17 + 4$	Is it even or odd?	How many digits does it have in decimal form?
101101001_2	How many digits does it have in binary form?	How many digits does it have in decimal form?

Consider an algebraic example: $x^2 + 4x - 5$; one can *see* that its graph cuts the x-axis at the point (0, -5). When it is expressed as $(x + 2)^2 - 9$, its minimum value of -9 becomes apparent. Students often fail to *read* such meanings into algebraic expressions other than thinking about substitution. This is an important algebraic skill to develop.

Effective teaching should attend to different *modes* of representation. NCTM (2000) includes the *representation* standard as part of the curriculum. It stated:

> Instructional programs from prekindergarten through grade 12 should enable all students to –
>
> - create and use representations to organize, record, and communicate mathematical ideas;
> - select, apply, and translate among mathematical representations to solve problems;
> - use representations to model and interpret physical, social, and mathematical phenomena. (p. 67)

The following sections explore how multiple representations can be effectively used in mathematics lessons.

3.2 *Concrete → Pictorial → Abstract* (CPA)

Bruner (1964a) proposed three modes of (external) representation of mathematics called the *enactive, iconic,* and *symbolic* modes. In the 1980s, the Singapore curriculum team led by Kho (Teng, 2014) converted Bruner's theory into the teaching approach called CPA (Concrete → Pictorial → Abstract) and produced a set of textbooks and teaching materials in the *Primary Mathematics Project* (PMP) that utilised this approach. The PMP series was used from the mid-1980s to early 2000s and has now been adapted in several countries as exemplars of the so-called *Singapore Maths.* This approach is suitable for primary and lower secondary levels because it is consistent with students' cognitive development identified by several psychological theories. It also makes less demand on students' language competence because concepts are initially formed from acting on concrete materials rather than processing verbal explanations. Recent educators have used different labels for this approach, for example, the Concrete-Representational-Abstract (CRA) method (Flores, 2010; Sousa, 2008; Tapper, 2012).

During the Concrete stage, students should manipulate real and tangible objects that embody the mathematical idea rather than just watch teacher's demonstrations. The prime number activity in Figure 2.1 is an example of this. Different hands-on activities should be used to introduce the same concept. An alternative to the activity in Figure 2.1 is to get students to complete the factor trees of numbers and discuss the patterns.

Concrete experiences with paper folding, measuring, cutting, and so on, enable students to answer questions that may require such exposure. Consider the following item modified from the Singapore Primary School Leaving Examination (PSLE), a public examination taken by all Grade 6 Singapore students at the end of primary schooling:

A rectangular piece of paper is folded to form two symmetric parts as shown below (not drawn to scale). What is the area of the piece of paper before it was folded?

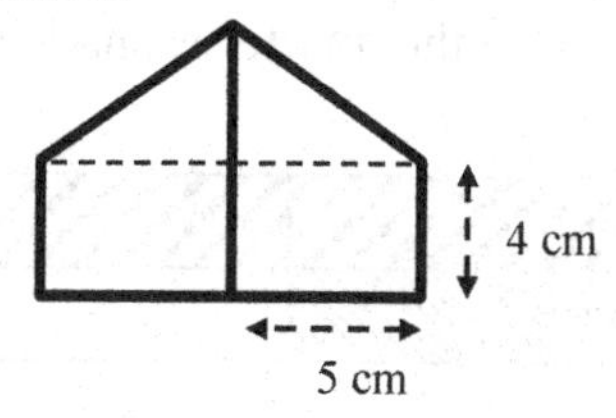

Figure 2.5. An item modified from PSLE

Students with hands-on experiences are able to visualise what happens in the above item and hence solve it successfully. Once this problem has been solved, students can pose their own questions by changing some of the numbers and checking their solutions by making a model of the situation.

There are two types of diagrams or pictures used in the Pictorial stage: *literal* ones that resemble the actual objects (e.g., a drawing of four apples to represent 4, as found in many K-1 books) and *iconic* ones that use mathematical shapes to represent objects (e.g., four dots to represent four apples). Iconic diagrams have special conventions that need to be taught. For example, the diagram in Figure 2.6 can represent a square, a square pyramid, or four right-angled triangles. The context of the problem should help students arrive at the correct interpretation, and this is a skill to be developed.

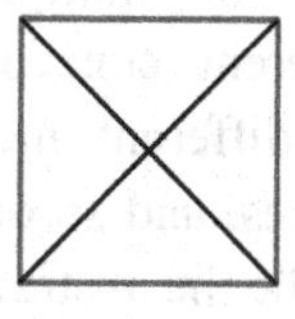

Figure 2.6. Same diagram with different meanings

Mathematical diagrams may depict the concrete activities used in the initial stage of concept formation. This use of diagrams helps students to better visualise the mathematical properties. Consider the addition of fractions, say ¼ + ½. A concrete approach is to hold a piece of paper lengthwise into four equal parts and shade one part. Then fold it by the width into two equal parts and shade one part. This action can then be represented by the diagram shown in Figure 2.7. In this diagram, the overlapping one-eighth from the quarter is shaded again.

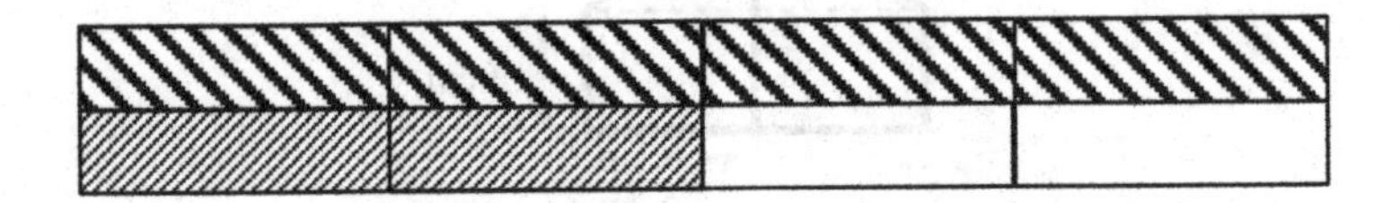

Figure 2.7. Diagram to reflect actions about fraction addition

This diagram closely reflects the hands-on actions, thereby strengthening students' visualisation of the process of fraction addition with different denominators. It also leads nicely to the abstract rule:

- The product of the denominators gives the total number of newly created equal parts. This also reinforces the concept of equivalent fractions.
- Counting the number of shaded parts gives the numerator of the sum.

The Abstract stage leads students to the mathematical representations, usually in words and symbols. Different symbols can used to represent the same idea; for example, multiplication can be shown as 2×3, $2 \bullet 3$ (the dot is above the base), $2y$ (and not $y2$), xy (same as yx, without the operation sign!), and $(x)(y)$ (see Chapter 3, Section 2.1 for an instructive vignette about the concept of multiplication). Conversely, the same symbol may convey different concepts or rules. For example, the addition symbol (+) has different meanings for numbers, algebraic expressions, vectors, matrices, and so on. Students are required to learn these different ways to handle the abstract mode.

Past experiences suggest that some teachers skip the Concrete stage, while other teachers keep the students too long at this stage by delaying the transition to the Pictorial and Abstract stages. Both practices are ineffective and inefficient use of limited curriculum time.

3.3 *Multi-Modal Strategy (MMS)*

At higher grade levels, the three CPA modes are inadequate to promote deep understanding and require refinement. In the 1990s, I worked with Brunei teachers to develop the *Multi-Modal Strategy* (MMS) by extending these three modes to six (e.g., Wong & Veloo, 1996; Wong, 1999). These modes are used to describe the same mathematical concept or rule.

MMS uses terms and format that are easier for teachers and students to use compared to a similar model by Lesh, Landau, and Hamilton (1983). Each of the six modes is associated with a key action verb. A brief explanation follows:

- *Real thing* or concrete material (*do*) that embodies the mathematical idea. Students must manipulate the objects rather than just observe teacher's demonstrations. These actions help to scaffold the meanings of the concepts and the reasoning underpinning the rules.
- *Word* (*communicate*), initially referring to the concrete experience, later to be written down. Correct mathematical language should be used. Students listen to teacher talking mathematics and then learn to do so themselves during class discussions or writing out mathematics solutions and journals about their learning.
- *Number* (*calculate*), which is obtained from counting or measuring; numbers are often used to verify algebraic rules.
- *Diagram* (*visualise*), which may be literal or iconic as explained earlier. Students develop visualisation by working with different types of diagrams.
- *Symbol* and notation (*manipulate*), the most abstract and precise form of the mathematical language.
- *Story* (*apply*), such as word problems as applications of mathematics to real-life and other contexts in mathematics, science, and so forth.

In general, the teaching of every concept or rule should cover these six modes. A recommended teaching sequence to cover these modes is shown in Figure 2.8. Notice that *virtual manipulative* has been added to this sequence in view of the recent drive in many countries to promote the use of computer-based learning; this will be discussed in Chapter 6. However, teachers should try different sequences of introducing the modes and note which sequence works well for different students and topics. Teachers can refer to these modes in their lessons to encourage students to use them as prompts to monitor their own learning as one form of metacognition.

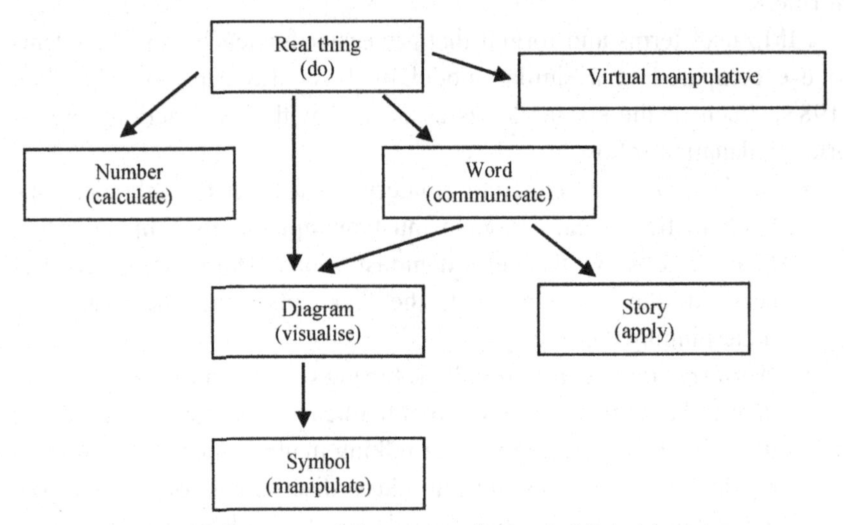

Figure 2.8. Recommended sequence of using multi-modal strategy

The above sequence highlights that several lessons are required to cover the six modes for the same concept or rule. A multi-modal strategy thinkboard, which is an extension of the four modes in Haylock's (1984) thinkboard, can be used to monitor the separate lessons and eventually to summarise what has been covered in separate lessons in a holistic view. This summary should be developed through whole class discussions, group work, or individual effort, rather than given in its final form to the students to be memorised or learned. This approach will help students see the connections among these different modes in their own ways. This highlights that there is no *final* thinkboard for any particular concept or skill. This active processing shows that the total, holistic view is more than the sum of the individual modes. This results in deeper understanding.

Figure 2.9 is an example of a completed multi-modal thinkboard for fraction division. The concept of fraction division used here is called the *measurement* (or *quotative*) model.

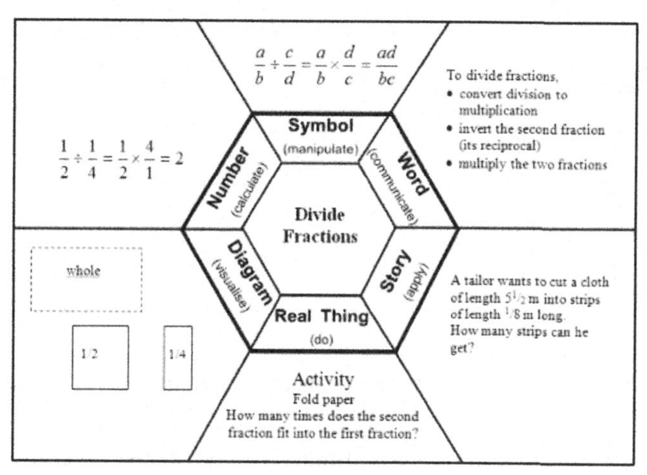

Figure 2.9. Multi-modal strategy thinkboard: fraction division

Other forms of summarising the six modes can be used in lieu of the hexagonal thinkboard. Figure 2.10 shows how a student had produced a multi-modal table for fraction division. He also added positive comments about this exercise, which one hopes will encourage more teachers to use this teaching tool.

The thinkboard and the multi-modal table can also be used to assess conceptual understanding. Give students one or two of the modes and ask them to complete the remaining modes using their own examples (Bishop, 1977). The ability to flexibly move from one mode to another in any order is an indicator of genuine understanding, as stated by Gardner (1991):

> Genuine understanding is most likely to emerge, and be apparent to others, if people possess a number of ways of representing knowledge of a concept or skill and can move readily back and forth among these forms of knowing. (p. 13)

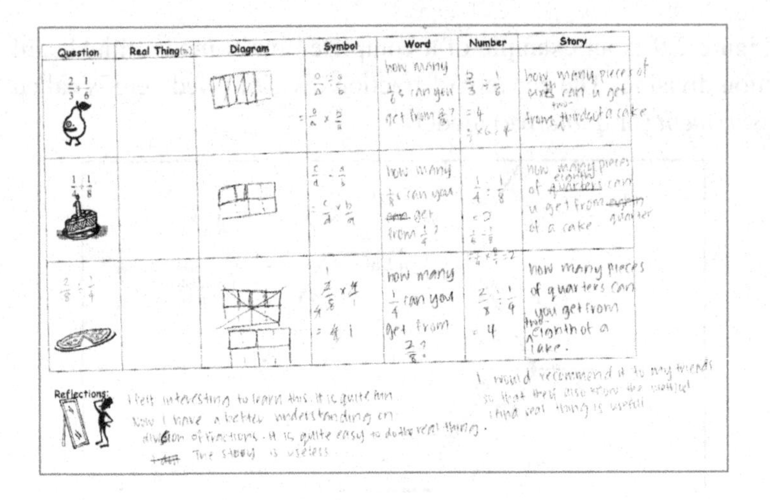

Figure 2.10. Multi-modal strategy in table form: fraction division

Liebeck (1984) proposed a similar 4-stage sequence comprising E (*experience*), L (spoken *language*), P (*pictures*), and S (written *symbols*). These different approaches to multiple representations are supported by well-known cognitive theories enunciated by Bruner, Piaget, Dienes, and other researchers. These approaches provide important guidelines for lesson planning.

3.4 *Modes of representation vs. modes of processing*

Of the five primary sensory channels, the following three are commonly used in mathematics learning (e.g., Arem, 2010):

- Tactile or kinaesthetic, through touch, gestures, and actions.
- Sound, auditory, or oral, involving speaking and listening.
- Sight or visual; this includes reading, writing, and drawing.

The multiple modes of representing mathematics can be linked to the predominant modes of processing, even though the match is not exact. This is suggested in Table 2.3. In this table, *write* refers to writing out mathematical texts and solutions by hand rather than typing them out on the computer.

Table 2.3

Modes of representation vs. modes of processing

Modes of Representation	**Kinaesthetic**	**Auditory**	**Visual**
Real thing	touch, action		
Word	gesture	read aloud, listen	read (silent), write
Symbol		read aloud, listen	read (silent), write
Number		read aloud, listen	read (silent), write
Diagram			draw
Story		read aloud, listen	read (silent), write

Students may begin to work on an activity in a mode they prefer (*learning preference* or *learning style*), but their learning should be extended beyond their preferred mode or comfort zone in order to deepen their learning. A recent review of research by Coe, Aloisi, Higgins, and Major (2014) concluded that "there are no benefits for learning from trying to present information to learners in their preferred learning style" (p. 24). This cautions against the popular belief that students learn better if the activities match their preferred learning style. According to neuroscience, different modes of processing are linked to different parts of the brain: auditory to temporal lobe, visual to the occipital lobe, and kinaesthetic to the parietal lobe. Getting students to use all the modes of processing through the multi-modal strategy is likely to help them to literally *grow* their brain by establishing new neural pathways.

Similarly, teachers have their own preferred mode of teaching, and this is likely to be predominantly verbal, through a lot of talking and explaining. They should also widen their own teaching styles to accommodate the different modes of their students.

4 Conceptual Connections

Conceptual understanding or conceptual knowledge involves seeing connections among concepts rather than treating them as isolated ideas. Hiebert and Lefevre (1986) provided this succinct explanation:

> Conceptual knowledge is characterized most clearly as knowledge that is rich in relationships. It can be thought of as a connected web of knowledge, a network in which the linking relationships are as prominent as the discrete pieces of information. Relationships pervade the individual facts and propositions so that all pieces of information are linked to some network. (pp. 3-4)

Connections among concepts are seldom made explicit in lessons or textbooks. These connections can be communicated to students using a variety of graphic organisers: *knowledge map, mind map, thinking map, cognitive map, semantic network, Venn diagram, tree diagram, Caroll diagram,* and *concept map* (see Jin, 2013, for a comparison of these maps). These graphic organisers can be used in different ways:

- As an advance organiser (Ausubel, 1968) to introduce a new topic; see Chapter 9.
- As a plan to show the progress as the class works through the topic in one or more lessons.
- As homework for students to work individually or in groups to summarise what they have learned about one or more topics, focussing on connections rather than details of rules.

Four types of graphic organisers are discussed below.

4.1 *Carroll diagram*

Carroll diagram, named after Lewis Carroll[6], is a two-way table with different attributes along the rows and columns so that examples that satisfy combinations of these attributes are inserted into the respective cells. It is a simple yet powerful tool to classify objects based on whether

---

[6] http://en.wikipedia.org/wiki/Carroll_diagram

or not they satisfy the properties of two concepts. Table 2.4 shows two Carroll diagrams with some examples; students can add more examples to the table to show their understanding of the respective concepts.

Table 2.4
Carroll diagrams: (a) Prime numbers (b) Perfect squares

(a)	**Prime**	**Composite**
Even	2	4, 6
Odd	3, 5	9, 15

(b)	**Perfect square**	**Not perfect square**
Multiple of 5	25, 625	15
Not multiple of 5	4, 49	123

4.2 *Venn diagram*

This is normally covered in the topic about *sets* to show subset-set relationship: set A is a subset of set B if every element of A is also an element of B. Intuitively, set B is a *larger* set, and this idea is loosely applied when Venn diagrams are used to depict conceptual relationships. In Figure 2.11(a), it is obvious that the set of natural numbers is a subset of the set of integers because the latter contains the former and includes additional elements such as negative numbers.

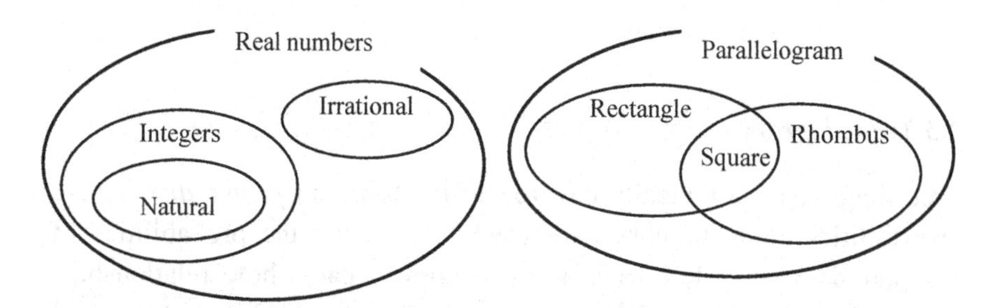

Figure 2.11. Venn diagrams. (a) Types of numbers (b) Types of parallelograms

In Figure 2.11(b), however, many students fail to understand that the set of squares is contained within the set of rectangles. They seem to focus on the idea that a square has *more* properties than a rectangle (e.g., four equal sides for a square but not for a rectangle), and by drawing a parallel with integers in the case of Figure 2.11(a), they think that the set of rectangles should be included in the set of squares.

The difficulty of understanding why "a square is a rectangle" arises because, according to van Hieles' theory (1986), the students are in transition between level 2 (Analysis) and level 3 (Relational) of geometry thinking. To help these students grasp the logic of the argument, try the following steps:

a) Explore the meaning of everyday statements, such as "a rose is a flower but a flower is not always a rose" or "*all* roses are flowers but not *all* flowers are roses."
b) Explain that in the above case, *flower* is a more general (or *super-ordinate*) concept than *rose* (a *sub-ordinate* concept).
c) Following from (b), ask students to list all the properties of a square and of a rectangle and note that a square has *all* the properties of a rectangle, so *rectangle* is a more general concept than *square*.
d) Now, apply the same everyday language use to help students arrive at the correct conclusion.
e) To reinforce this learning, ask the students to work with similar statements, such as "Is a rhombus a rectangle?" or "Is an integer a natural number?" To arrive at the correct answers, teachers and students must begin with the same definitions of these mathematical objects.

4.3 *Tree diagram*

Tree diagrams are typically covered in the topic on *probability,* where probabilities along branches are used to compute the probabilities of compound events. Here, it is used to convey part-whole relationships among concepts. Figure 2.12 shows a tree diagram of different types of quadrilaterals, with the super-ordinate concept *quadrilateral* at the top. The shapes are arranged according to the following definitions:

- Trapezium: only one pair of parallel sides.
- Trapezoid: no parallel sides.
- Kite: convex quadrilateral with two pairs of equal adjacent sides; note that a kite may or may not be a trapezoid.
- Rhomboid: a parallelogram with unequal adjacent sides.
- Deltoid: non-convex quadrilateral with two pairs of equal adjacent sides.

As mentioned in Section 2.1, differences in how these geometry concepts are defined can lead to different tree diagrams. See Usiskin and Griffin (2008) for alternative definitions of these objects.

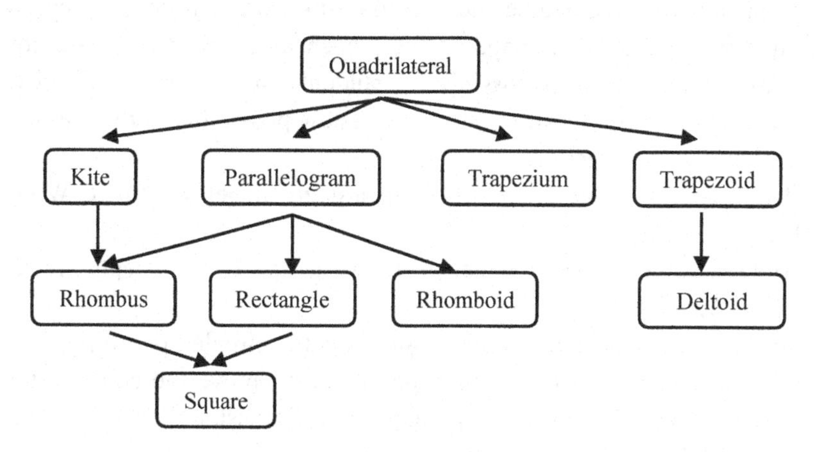

Figure 2.12. Types of quadrilaterals

Although Venn diagrams, tree diagrams, and Carroll diagrams (to a lesser extent), are used in textbooks, research about their effects on learning and assessment is limited. In contrast, concept maps have attracted much attention among mathematics educators; see the extensive review in Jin (2013). A broad survey of concept maps is given in the next section.

4.4 *Concept maps*

Concept maps look like tree diagrams but not all the concepts in a concept map have to satisfy the part-whole hierarchy. This tool has a relatively long history in science education since the early 1980s, when Novak and others (see Novak & Gowin, 1984; Novak, 2010) used it to promote Ausubel's cognitive theory of meaningful learning in science. Over the past three decades, mathematics educators who have investigated the effects of concept maps generally recommend its use as a learning-teaching strategy, and it is now included in pedagogy texts (Afamasaga-Fuata'I, 2009; Krpan, 2013) and curriculum documents. The Singapore mathematics curriculum recommends that students use concept maps to summarise their learning as part of reflective review. Computer-based concept-mapping tools such as *IHMC Cmap*[7] are now available, which make it possible for students to collaboratively create concept maps across locations and time, a new possibility not yet utilised in mathematics instruction.

The steps to create a concept map are given below (Jin & Wong, 2011; Krpan, 2013):

a) Determine a list of concepts to work on. The concepts are the *nodes* of the map.
b) Enclose each concept in a box, usually circular or elliptic. For convenience, write the concepts on cards, move the cards around to the intended positions, perhaps after discussion with others, then copy the final map onto paper.
c) Place the super-ordinate concepts near the top of the page so that the other concepts can be arranged in a hierarchical way below them. However, hierarchy may be optional for students at lower grade levels.
d) Work on one concept at a time and place it close to related concepts in a more-or-less hierarchical way.
e) Connect related concepts by lines. These lines may be directional with arrows or non-directional without arrows. If appropriate, use

---

[7] http://cmap.ihmc.us/

lines of different thickness to indicate strength or importance of the relationships.

f) Write linking phrases alongside the connecting lines to describe the nature of the relationship. Common linking phrases include *is an example of, has,* and *can be divided into*. These can be converted into meaningful statements called *propositions;* see the example in Figure 2.13.

g) Use cross-links to connect pairs of concepts that appear in different parts of the map. In some cases, the concepts may be re-arranged to reduce weak cross-links.

Figure 2.13 shows part of a concept map about elementary algebra. The following propositions can be generated, and examples are attached to these propositions:

- An algebraic term is a product of constant and variable; e.g., $2x$, $5xyz$ (could include more than one variable in a single term).
- An algebraic expression is a sum of algebraic terms, e.g., $2x - 5xyz$.

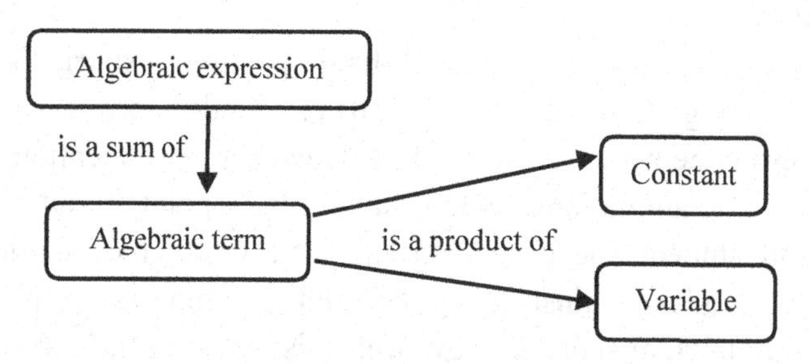

Figure 2.13. A concept map about algebraic expression

Entrekin (1992) explained how to construct a concept map at the beginning of a lesson to help students review previously learned materials. Ask someone in the class to give a concept about the topic, and write it on the board. Ask another student to think of an associated concept and how it is related to the first concept. Continue this process to build the map with the class. She did not use linking phrases in her map, but found that some students drew their own "small maps" in the corner of a test to help them recall the materials.

Concept maps have been used to assess conceptual understanding. They can evaluate more directly the connections among concepts held by the students compared to traditional problem solving tasks. Different schemes have been designed to score the quality of student-generated concept maps. Novak and Gowin (1984) considered the validity of propositions, hierarchy, cross links, and examples. McClure, Sonak, and Suen (1999) included scoring against expert maps. Jin (2013) scored the number of links and the quality of the linking phrases. She also applied social network analysis to analyse these scores, resulting in scores for individual concept maps and the quality of a *collective map* of the whole class. These concept map scores are found to have concurrent validity with other assessment tasks, such as a task assessing definition-examples-non-examples and typical problem solving (Jin, 2013). For classroom use, teachers should focus on giving descriptive feedback, such as pointing out misconceptions, missing links, and extensions (Krpan, 2013), rather than scoring the concept maps using sophisticated scoring schemes. Furthermore, concept maps created by students before and after instruction can be used to determine changes in their conceptual understanding.

There is evidence that students have positive experiences using concept mapping (Jin & Wong, 2011). Under proper guidance, constructing concept maps can be fun. However, in education systems that focus on examinations, Nannestad (1998) pointed out that both teachers and students need to overcome the conflict of getting good examination grades against using concept mapping as a process of learning without rigid scoring. How well this tension can be resolved by the teacher and students will affect the acceptance of this strategy in mathematics teaching, learning, and assessment.

5 Concept Questions

To promote conceptual understanding, teachers should ask the following *concept questions* in lessons during question-and-answer interactions or discussions. They can also set them as prompts for students to write reflective review about the lessons. These questions should cover the materials discussed in earlier sections, in particular about *meanings, examples, non-examples, modes of representation,* and *connections*. They include:

- What does ... mean in ...?
- What do you understand by ...?
- What is the definition of ...?
- What is an example of ...? What is another example of ...?
- What is a non-example of ...? What is another non-example of ...?
- How can you show ... on a diagram?
- Tell a story of how ... is used?
- How is ... related/connected/linked to ...?

Students should also be trained to ask themselves similar questions in order to monitor how well they have learned the concepts. This is an important component of metacognition. Subsequent chapters will also include similar reflection questions for the respective curriculum components. This innovative approach is offered in this book as a consistent way to inculcate metacognition among students in their mathematics learning.

6 Concluding Remarks

Students need to understand the concepts underpinning the problems they are asked to solve. This chapter has dealt with the essential features of concept formation and concept mastery that need to be included in the mathematics curriculum and delivered effectively in lessons. Effective techniques include CPA, multi-modal strategy and processing, and graphic organisers. However, traditional mathematics problems do not specifically ask about the concepts; instead, they require mastery of skills or rules that are associated with the concepts. The next chapter will cover these skills so that they can be applied directly to solve mathematics problems.

Chapter 3

Skills: Use Rules Efficiently

Mathematics skills must be mastered to high degree of accuracy and speed in order to be applied efficiently to solve various types of problems. Direct instruction has been found to be more effective than unguided discovery to help students build these skills. This chapter examines how direct instruction can be better implemented by including alternative explanations, addressing student mistakes, covering a variety of worked examples, and paying attention to the cognitive loads of problems. These changes should be included in effective lessons to ensure skill mastery.

In order to master level N, *you must have internalized level* N – 1.
Steven Krantz (US mathematician)

1 Nature of Mathematical Skills

The term *skills* is used here to refer to the ability to perform rules, procedures, processes, techniques, algorithms, and methods fluently to high degree of accuracy and speed. Some rules may be construed as *tricks* or *shortcuts,* such as "bring over and change sign" or *BODMAS* (Brackets, Order, Divide, Multiply, Add, Subtract)[8], but almost all mathematical rules are not tricks, because they are based on fundamental concepts. They are step-by-step instructions to arrive at the right answers. Within the school context, students are also expected to be able to apply these rules with speed within stipulated times, without constantly referring to explicit instructions, worked examples, or formula

---

[8] http://www.math-only-math.com/bodmas-rule.html

lists, showing that they have internalised the skills. In essence, skills are the *know how* of mathematics.

1.1 *Alternative procedures*

Alternative procedures are often available to solve the same problem, but they are not equally efficient or can be readily understood by the target students. Consider the procedures used to find the highest common factor (HCF) of two numbers, say 36 and 60. This builds on the concept of HCF we encountered in Figure 2.5.

Method 1: Definition
Apply the definition of HCF: list the factors of each number, then determine the largest common factor.
Factors of 36: 1, 2, 3, 4, 6, 9, 12, 18, 36
Factors of 60: 1, 2, 3, 4, 5, 6, 10, 12, 15, 20, 30, 60
Common factors of 36 and 60: 1, 2, 3, 4, 6, 12
The HCF of 36 and 60 is12.

Method 2: Prime factorisation
Express each number in its prime factorisation. For each prime factor, choose the one with the lower power and multiply all the common prime factors together.
Prime factorisation of 36: $2^2 \times 3^2$
Prime factorisation of 60: $2^2 \times 3 \times 5$
The HCF of 36 and 60 is $2^2 \times 3 = 12$.

Method 3: "Staircase" or division method

Divide both numbers by the smallest possible prime factor.
Repeat the process on the resulting quotients until there is no more common prime factor.
Multiply the prime factors so determined.

2	36	60
2	18	30
3	9	15
	3	5

This is shown on the right, and the HCF is $2 \times 2 \times 3 = 12$.

Method 4: Euclidean algorithm
This ancient algorithm (ca 300 BC) is still used as an efficient method to determine the HCF of two numbers, but it is seldom taught in schools nowadays.
Divide the larger number by the smaller number (divisor) and write out the result: $60 = 36 \times 1 + 24$; 24 is the remainder.
Divide the divisor by the remainder: $36 = 24 \times 1 + 12$.
Repeat the process by dividing the first remainder by the second remainder until the remainder is zero: $24 = 12 \times 2 + 0$.
The HCF is 12.

From the cognitive theory of meaningful learning, Method 1 should be used, but it is not efficient. Method 3 is most efficient and quite popular, but many students do not understand why it works. Students who know only this staircase method will have difficulty solving problems similar to this one:

The HCF of 18, 30, and x is 6. Find the possible values of x given that it is less than 40.

Method 2 is fairly efficient and meaningful, but some students are puzzled over why the lower powers for the common prime factors are chosen. Method 4 can be easily programmed to run on a computer or a spreadsheet. For the sake of promoting versatile problem solving, it is advisable to teach alternative procedures and let students decide on the methods they prefer to use. Obviously, more curriculum time is needed but the extra time is worthwhile because the exposure to alternative methods can promote flexible mathematical thinking. After the students have found the answer in any of the above methods, they must remember to check that it is indeed a factor of the given numbers.

1.2 *Conditions for procedures*

Many mathematical procedures require certain conditions to be satisfied before they can be used. For examples, Pythagoras' Theorem applies to only right-angled triangles, the multiplication of probabilities requires that the events are independent, and the median is determined only after the values have been arranged in ascending or descending order. Thus, before using these rules, students must check that the necessary conditions have been satisfied. Unfortunately, many of them have yet to develop this desirable mathematical habit. The following example taken from Galbraith's vitality test (Galbraith, 1982) illustrates this shortcoming among pre-service teachers.

Find the derivative of $f(x) = \sqrt{2x - x^2 - 1}$.

About 84% of Singapore and 62% of Australian pre-service teachers (see Chapter 2, Section 2.1) could compute the derivative correctly, which is $\frac{1-x}{f(x)}$. But they did not realise that the derivative does not exist because the given function is defined for only the single point (1, 0). Given that a sizeable number of pre-service teachers did not check the conditions before they applied the rules, it is important to inculcate this habit as early as possible in the student's journey through mathematics.

1.3 *Hierarchies of mathematical skills*

Similar to concepts, mathematics skills can also be arranged in hierarchies, and the following three criteria can be used to build these hierarchies.

Criterion 1. Logical structures

The skills can be arranged according to their logical structures. One may begin with a topic and work in a top-down way to break it down into pre-requisite skills at lower levels. This is shown in the two examples below:

- Multiplication of numbers follows from addition because the former can be defined as repeated addition.
- Fraction division comes after fraction multiplication because of the *invert and multiple* rule.

Criterion 2. Professional knowledge
Teachers' professional knowledge of what works for their students and for which topics can be used to design the hierarchies. This assumes that the teachers have reflected well on their teaching and shared their experiences with colleagues. The topic sequences derived in this experiential way usually align well with the logical ones.

Criterion 3. Research
Many researchers have gathered quantitative and qualitative data about student mathematical solutions and errors (e.g., Hart, 1981). These analyses often lead to hierarchies of skills based on facility levels or item difficulties. The *Quantile Framework for Mathematics* [9] is a tool developed by MetaMetrics in the US that places mathematics skills and concepts along scales called *QSC* (Quantile Skills and Concepts). These QSCs describe the difficulty in learning these skills and concepts within a topic as well as across different topics. A sample of QSC sequence is given in Table 3.1.

Table 3.1
Sample of Quantile Skills and Concepts (QSC)

Quantile Measures	Mathematics Skills and Concepts
160Q	Identify and name basic solid figures
300Q	Make different sets of coins with equivalent values
410Q	Round whole numbers to a given place value
780Q	Write numbers using prime factorization
850Q	Describe data using the mean
1200Q	Graph exponential functions of the form $f(x) = ab^x$

An advantage of the QSC scheme is that the measures can be used to sequence skills across different topics. In Table 3.1, skills in basic geometry are compared to those about primes, basic statistics, and algebraic functions. A useful exercise is to compare the sequences of topics in the intended curriculum documents with the Quantile measures.

[9] https://www.quantiles.com/

This will enhance teachers' understanding of the curriculum, and this knowledge can be used to prepare quality schemes of work and lesson plans.

The hierarchies are used to assign topics to grade levels in the intended curriculum. An algebra example is shown in Figure 3.1. Solution of linear equations appears in Grade 7 and solution of quadratic equations in Grade 8. Lessons are planned to move from the lowest level upward.

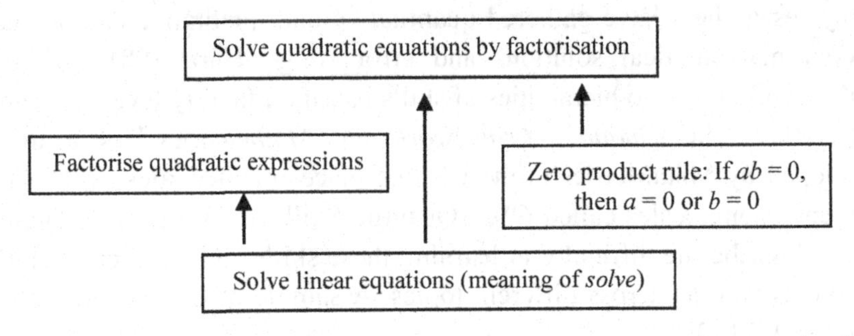

Figure 3.1. A simple hierarchy of algebraic factorisation

2 Skills vs. Concepts

We noted in Table 2.1 of Chapter 2 that skills and concepts are intrinsically connected to one another. However, it is important to make a distinction between them because understanding concepts and mastering skills require different cognitive processes. Confusion between concepts and skills may lead to poor teaching.

2.1 *An example from fraction division*

Consider fraction division such as ½ ÷ ¼, as captured in Figure 2.9 of Chapter 2. The *Concept* aspect is to see this operation as finding how many quarters (the divisor) can fit into a half (the dividend); this is called the *measurement* model. For this division, the concept of *equal sharing* (e.g., share a half equally among a quarter people) does not make sense. The *Skill* aspect refers to the technique used to give the correct answer without paying too much attention to its meaning. The most common technique taught in schools is the *invert and multiply* rule, i.e., $\frac{1}{2} \div \frac{1}{4} = \frac{1}{2} \times {}^{4}/_{1} = 2$. The *Process* aspect (to be discussed in Chapter 4) is to justify this rule by referring back to the concept. When students make mistakes in applying the rule to complex fractions such as ${}^{3}/_{5} \div {}^{7}/_{10}$, some teachers may conclude that the students do not *understand* the *concept* of fraction division, whereas the actual problem may lie in computational fluency. This confusion may lead these teachers to repeat the same explanation about concepts and give inappropriate remediation. In a similar vein, when students say that they "understand the concept but cannot do the computations," they probably know neither (Ellenberg, 2014). The more discerning teachers will be able to identify the real problem and provide the appropriate follow-up.

2.2 *Procept*

From a theoretical perspective, Tall and Vinner (1981) introduced the construct *concept image* to include the mental images, properties, and processes associated with a concept, and they found differences between concept images held by students and concept definitions which may be personally created or mathematically given. Later on, Gray and Tall (1994) introduced the term *procept* to stress that a concept and its associated procedure may be treated as an entity and this amalgam is then represented by symbols. Both constructs are helpful in drawing teacher's attention to these differences when the aim of instruction is to promote versatile mathematical thinking.

As the first example, the procept of a fraction, symbolised by ${}^{a}/_{b}$ consists of the part-whole concept (a *fraction*) and a division process ($a \div b$). However, as pointed out by Thomas (2008), about 22% of 14 year-old students in UK did not consider these two as equivalent, suggesting that they had not been able to combine an object and its related process.

As a second example, the procept of an algebraic expression, say $2x + 3y$, is a sum of terms and the process of substituting values to determine a specific value; see the concept map in Figure 2.13 of Chapter 2.

2.3 *Limit and recurring decimals*

The third example of procept is more complex than the above two, and it is about the concept image of *limit* in calculus and number sequences. This concept image can be explicated as: a sequence of numbers *approaches* but never *reaches* the limit. This is applied now to the persistent learning problem about recurring (repeating) decimals.

This problem has been widely discussed by educators, including Davis, Ely, Schwarzenberger, Tall, Thomas, and Vinner. Many students and teachers believe that 0.999... can be made as close as possible to 1 but it is never equal to 1. On the other hand, they readily agree that 0.333... is equal to $^1/_3$, and the notion of 0.333... getting close to but not equal to $^1/_3$ does not seem to arise. From the perspective of procept, the concept for a recurring decimal is linked to the process of generating it by fraction division. The fraction $^1/_3$ can be divided out to show that it recurs. This is also the case with $^1/_9$, $^2/_9$, and so on until $^8/_9$. Now, this is not possible to obtain a recurring decimal with $^9/_9$, and the students and teachers fail to link this fraction to 0.999.... This shows that the procept is incomplete for this recurring decimal.

To convince students that 0.999... is equal to 1, the following formal *proof* is frequently offered:

Let $S = 0.9 + 0.09 + 0.009 + \dots$

Then $10S = 9 + 0.9 + 0.09 + \dots = 9 + S$, so $S = 1$.

My experience with teachers is that they agree that this proof works, but many of them still do not change their original belief. As pointed out by Ellenberg (2014), the above multiplication method can lead to the strange result:

Let $T = 1 + 2 + 4 + 8 + \dots$

Then $2T = 2 + 4 + 8 + \dots = T - 1$, so $T = -1$.

The proper mathematical approach is to *define* the meaning of an infinite series in terms of its *limit* if it exists. In the case of S, prove that it is convergent and the limit is 1. This limit gives the meaning of the recurring decimal. On the other hand, T is divergent and does not have a limit, so the multiplication method does not work. This rigorous approach may be beyond most school students, but the teachers should still explain that the multiplication method does not work when the series is divergent.

Students often come to the wrong conclusion about whether a rational number is recurring or terminating by looking at the output from a calculator or spreadsheet such as *Excel* (up to only 30 digits). For example, $^1/_{29}$ = 0. 0344827586206896551724137931.... Its repetend (03...31) has 28 digits, and these are not displayed in the calculator or *Excel* output. As a result of this, some students conclude that this fraction does not recurring. One way to handle this is to write a simple division algorithm in *Excel* to print out as many of the decimals as needed; this will be shown in Figure 6.4 of Chapter 6.

Concepts and skills contribute differently to the ability to solve problems, and the ways to teach them also vary. However, the theory of procept suggests that the two components need to be integrated into a coherent complex to strengthen rule applications and mathematical thinking.

3 Direct Instruction: An Overview

Direct instruction (DI) has many different labels: *explicit instruction, expository teaching, direct teaching, teacher-directed learning, explicit teaching, chalk-and-talk, scripted lessons,* etc. It is deceptively straightforward: explain ideas and worked examples clearly, let students practise the rules, correct their work, and test them on similar problems. It might have its origin as one-to-one instruction between master and apprentice or even between parent and child. As a formal instructional strategy in the West, the essential features mentioned above are strongly supported by behaviourist theories and research by Thorndike, Skinner, Gagné, and others.

Beginning in the 1970s, DI has been derided as rote, mindless, and *drill and kill,* by advocates of the child-centred and discovery approach based on constructivism (e.g., Anderson, Reder, & Simon, 1999; Davis, Maher, & Noddings, 1990), but it is still the predominant teaching method for mathematics in many countries. This has created controversy among educators for the past four decades. One controversy includes the misconceived notion that learning through DI is necessarily passive and rote. In more recent years, however, there is greater acceptance of DI based on new research, meta-analysis, and a less rigid adherence to dichotomous views about learning. Reviews by educators such as Alfieri, Brooks, Aldrich, and Tenenbaum (2011), Coe, Aloisi, Higgins, and Major (2014), and Hattie (2012), conclude that DI is more effective than unassisted discovery, especially for struggling mathematics students. The authors of the most recent review of the Australian curriculum cited evidence to show "that explicit teaching is more effective than many other theories of teaching and learning" (Donnelly and Wiltshire, 2014, p. 125). They recommended that

> effective teachers employ a range of often different models of teaching and learning, depending on what is being taught, the ability and motivation of students, the year level and the nature of the intended outcomes. (p. 246)

For mathematics instruction, the US National Mathematics Advisory Panel (2008) reported that it found only eight quality studies comparing teacher-directed and student-centred instruction. Given this limitation, they concluded that "[h]igh-quality research does not support the exclusive use of either approach" (p. 45).

Many teachers, especially those from the East, are more competent in DI than in constructivist approaches. There is also strong interest in how mathematics is taught in the so-called *Confucian-Heritage-Culture* (CHC) of China, Hong Kong, and Taiwan (e.g., Fan, Wong, Cai, & Li, 2004). DI as practised under CHC has also been acknowledged as a strong factor of high performance of Shanghai students in PISA. This idea has now spread to Western countries. In late 2014, the England-China Project[10] organised an exchange of English and Shanghai primary mathematics teachers, and an interim report[11] concluded that English primary students made progress under the Shanghai method which includes the following features of DI:

[10] http://www.mathshubs.org.uk/what-maths-hubs-are-doing/england-china/;

[11] https://www.ncetm.org.uk/news/46090

- meticulously scripted lessons including follow-up questions to students' responses,
- constant reference to textbooks,
- emphasis on students using precise mathematical language to describe daily events, and
- approaching a topic from different angles.

DI and constructivist approaches have their own spheres of applications in teaching and learning. Furthermore, definitions of DI, constructivism, and student-centred learning have overlapping ideas. In a recent study of highly regarded mathematics teachers in the US, the authors (Walters, Smith, Leinwand, Surr, Stein, & Bailey, 2014) included in their definition of student-centred instruction features such as sense-making, building on prior learning, and strong relationships between and among the teacher and students. These are also essential features of quality DI.

In this book, I will focus on DI as the main approach to develop skill mastery, and in the next chapter on *Processes*, I will explore a combination of DI and the discovery approach to show how these two methods can complement each other to help students develop versatile mathematical reasoning.

4 Frameworks of Direct Instruction

Many frameworks of DI have been designed to focus on its different key features. Table 3.2 compares these key features under Gagné *Nine Events of Instruction* (Gagné, Briggs, & Wager, 1992) and Rosenshine's *10 Principles of Instruction* (2012). The alignments are not perfect, but the critical notion that most of these instructional events are teacher directed is quite apparent in both frameworks. The instructional events listed under these frameworks are often included in mathematics lessons; see Chapter 9.

Table 3.2
Frameworks of DI: Gagné and Rosenshine

Steps	Gagné	Rosenshine
1.	Gain attention of the students	
2.	Inform students of the objectives	
3.	Stimulate recall of prior learning	1. Begin a lesson with a short review of previous learning
4.	Present the content	2. Present new material in small steps, with student practice after each step 3. Ask a large number of questions and check the responses of all students
5.	Provide learning guidance	4. Provide students with models and worked examples to help them learn to solve problems faster 8. Provide scaffolds for difficult tasks
6.	Elicit performance	5. Guide student practice 9. Require and monitor independent practice
7.	Provide feedback	6. Check for student understanding
8.	Assess performance	7. Obtain a high success rate
9.	Enhance retention and transfer to the job	10. Engage students in weekly and monthly review

5 Telling and Explaining

Part of the oft-repeated comment of William Arthur Ward is: "The mediocre teacher tells. The good teacher explains."[12] To *tell* the class something is to mention the result or to show step-by-step how to complete a rule, without further elaboration. Skemp (1976) referred to this as *instrumental understanding*. To *explain* the same thing, however, the teacher will start from the students' current understanding and lead them to the new learning through demonstration with different modes of representation and question-and-answer. The distinction is shown in the following example about the decimal multiplication, 0.6×3.

[12] http://www.brainyquote.com/quotes/quotes/w/williamart103463.html

Tell — To calculate 0.6×3, first compute $6 \times 3 = 18$. Then count the number of decimals in the numbers, in this case, one. The answer has one decimal place, i.e., 1.8.

Explain — Recall the meaning of multiplication as repeated addition. So, $0.6 \times 3 = 0.6 + 0.6 + 0.6 = 1.8$, applying what we have learned how adding decimals.

We can show this multiplication in Figure 3.2. First shade 0.6 as shown. Multiplication by 3 means we have to shade two more sets to make 3 sets altogether; teacher to carry out the shading, which is not shown here. Now we have 1.8.

At the end of this demonstration, pose this question: the answer is smaller than 3 but bigger than 0.6. Why? (See a similar item in Figure 1.4, used to test MPCK of future teachers)

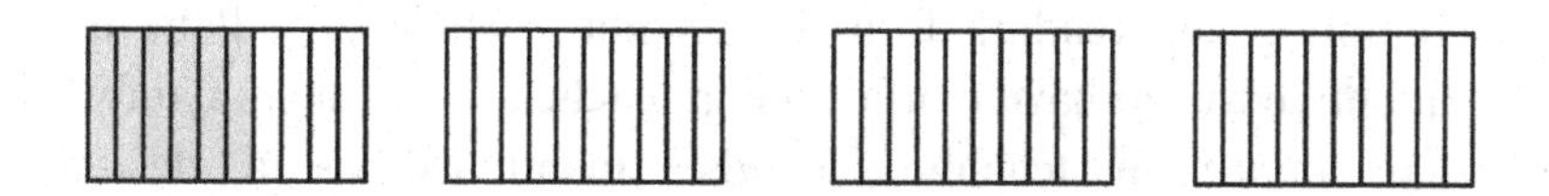

Figure 3.2. Decimal multiplication (incomplete shading)

Research has shown that teacher clarity in explaining concepts, rules, instructions, and many other things in class, ranks among the highly desirable teacher characteristics sought after by students (e.g., Lim & Wong, 1989; Wong, Kaur, Koay, & Jamilah, 2009). When students were asked to draw picture about their *best* mathematics teacher, many diagrams were about DI with neat work shown on the board. An example is shown in Figure 3.3.

Figure 3.3. A drawing about my *best* mathematics teacher

Teacher clarity can be developed through much deliberate practice of the following ways. These should be inculcated from pre-service training.

- Rehearse or memorise the difficult parts of an explanation to avoid fumbling and reading from lesson notes; students will be more attentive to and have confidence in teachers who can explain well. Confidence in teaching is also mentioned as a desirable characteristic of good teachers.
- Be enthusiastic in teaching, yet another important quality to possess. This will make explanations memorable to the students.
- Use the board effectively. Divide the board into separate columns and put up work systematically, so that by the end of an episode of the lesson, students can see a holistic picture of that episode. Pause for students to take notes, if necessary. Students should not listen and copy notes at the same time. Check that they have correctly copied the notes into the worksheet or notebook.
- Unless there is a need to show special materials such as intricate graphs that cannot be readily drawn on the board, avoid projecting step-by-step solutions using PowerPoint slides because this reduces interactivity with the class (see Tufte, 2003). Instead, combine explanation with question-and-answer to engage students

in thinking along with the explanation and to probe their understanding.

- Keep each explanation to about 10 minutes because many students quickly lose concentration and interest in long-winded explanations. Intersperse long explanations with engaging activities.
- A common teaching shortcoming is to explain more than one new rule without practice in between. It is more effective to use this sequence: *explain rule 1 → practise rule 1 → explain rule 2 → practise rule 2* and so on.
- Make effective use of body language, voice, eye contact, and gestures. In recent years, a few studies have investigates the so-called *gesture-speech synergy* (Goldin-Meadow, 2003). In mathematics teaching, gesturing that matches verbal explanations "actually takes less cognitive effort than giving an explanation without gesturing" (ibid, p. 166). For examples, one might raise a hand vertically when talking about height, or put two hands in a V shape when explaining a minimum point of a curve. These seem to be helpful.
- Add humour, stories from the history of mathematics, and personal mathematics experiences to the explanations.

6 Worked Examples

Worked examples are essential to help students learn how to complete specific rules and apply them in different contexts. They are used all the times by teachers in teaching rules and problem solving. The principles used to teach concepts, such as correct mathematics, coherent set of graded examples and non-examples, multiple modes of representation, mathematical and perceptual variability, can be applied to design appropriate worked examples for skills.

6.1 *Correct mathematics and real-world information*

All mathematics worked examples and practice problems must be mathematically correct. In Table 3.3, items (a) to (e) have mathematical flaws. Information purportedly about the real world must be realistic; this counters the tendency to simply make up unrealistic quantities about daily objects and events to make them look real; see items (e) and (f).

Table 3.3
Sample of flawed items

No.	Items	Comments
a)	The surface area A of a sphere is directly proportional to the square of its radius r. If $A = 40$ when $r = 2$, find the value of A when $r = 3$.	The item tests a rule about proportionality. However, $A = 4\pi r^2$. When $r = 2$, $A \approx 50.3$.
b)	Solve the equation: $\frac{a^2+ab}{ab+b^2}$.	Should distinguish between *solve* and *simplify*.
c)	If $y^2 = 12 - x^2$ and $x = \frac{8}{y}$, find the value of $(x + y)^2$.	It ends up with a negative value, which is impossible for real numbers (assumed).
d)	Calculate all the angles, given that the area is 5 cm^2. 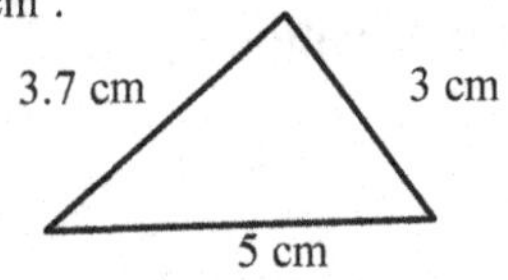	Heron's formula shows that the area should be 5.52 cm^2. The error may be small, but the teacher did not know of Heron's formula.
e)	The pendulum of a grandfather clock is 49 m long. The pendulum swings through an angle of 12^0. What is the arc length swept by the bob of the pendulum? ($\pi = \frac{22}{7}$)	Unrealistic length; the intention is to give a length that is multiple of 7 to simplify cancellation. Take $\pi = \frac{22}{7}$; or $\pi \approx \frac{22}{7}$; a single word can make a difference in terms of concept.
f)	The average January temperatures of some cities are given below: Brunei: 29^{0}C, Hong Kong: 17.5^{0}C, London: -2.5^{0}C, Moscow: -12.4^{0}C. What is the average January temperature of these four cities?	Meaningless computation.

6.2 *I do – We do – You do*

The "I do – We do – You do" sequence is a well-known, structured way to teach rules. This sequence can be integrated with Vygotsky's theory of transferring from external speech to private talk during learning (Manning & Payne, 1996). This is illustrated below with a set of worked examples for finding the lowest common multiple (LCM) of two or more numbers using prime factorisation. The main steps are given below:

a) Find the prime factorisation of each number.
b) For each prime factor, choose the one with the largest power. If a prime factor appears in one of the numbers but not in the other numbers, include it as well.
c) Multiple all the prime factors found in step (b). This gives the LCM.
d) Check that the answer obtained in step (c) can be divided evenly by all the given numbers. Note. This check shows only that the answer is a common multiple of the given numbers, but it does *not* check that it is always the lowest; an incomplete check is still useful to pick up careless computations.

I do: The teacher writes the above steps on the board and verbalises them aloud (external speech), while the students listen attentively.
Worked example 1: Find the LCM of 9 and 16.

We do: The teacher verbalises the steps while the students perform the steps accordingly; question-and-answer may be used here. A slightly more difficult example can be used.
Worked example 2. Find the LCM of 8, 9, and 12.

You do: Students practise the rule on similar examples in two main steps.
You do (1). Students murmur to themselves the steps while they carry them out; the teacher walks around to check seatwork.
Worked example 3. Find the LCM of 15 and 18.
You do (2). Students silently whisper the steps while they complete them. This becomes the private talk or internal dialogue that students may use subsequently on their own when they solve similar problems.

Worked example 4: Find the LCM of 6, 10, and 14.

The above sequence are followed by worked examples that are more challenging or involve realistic contexts; see the two problems below.

1. Find the smallest value of n such that the LCM of n and 6 is 24.
2. The lights on three lightships flash at regular intervals. The first light flashes every 18 seconds, the second every 24 seconds, and the third every 30 seconds. The three lights flash together at 2200 hour. At what time will they next flash together again?

6.3 *Cognitive Load Theory*

According to Sweller's Cognitive Load Theory (Sweller, 1992; Sweller, van Merrienboer, & Paas, 1998), three types of cognitive load are placed on the working memory when students learn new rules to solve problems.

- *Intrinsic* cognitive load: This depends on the nature of the task. The steps for LCM above exert an intrinsic cognitive load on the student. With practice and mastery, the intrinsic cognitive load would lessen.
- *Germane* cognitive load: This refers to the working memory required to consciously process the instructions in order to construct a schema about the task so that it can be stored in long-term-memory (LTM). Conscious processing includes self-explanation (e.g., private talk) and memorisation strategies. For example, through practice, the steps for finding LCM become a schema in LTM.
- *Extraneous* cognitive load: This depends on how the task is presented to the student, usually in a combination of words, symbols, and diagrams. The extraneous cognitive load may be lowered by using appropriate learning materials and instructional process. In the case of LCM, extraneous cognitive load will be less for single digit numbers than for multi-digit numbers. Referring to the intermediate steps (e.g., in worksheet or on the board) during problem solving will reduce this load.

The total cognitive load is the sum of these three types of cognitive load. According to Sweller, van Merrienboer, and Paas (1998), the most promising way to measure cognitive load is to ask the student to report their mental effort on the task using a uni-dimensional rating scale. They also cited studies in algebra, statistics, and programming, to support the counter-intuitive hypothesis that "worked examples facilitate learning and problem solving more than solving the equivalent problems" (p. 274). Part of the reason for this is that studying worked examples reduces extraneous cognitive load, compared to the means-ends analysis required during problem solving. This is known as the *worked example effect.* However, they also cautioned that students may be fixated on "stereotyped solution patterns that may inhibit the generation of new, creative solutions to problems" (p. 275); see Section 6.4 below. To counter this problem, they recommended five ways to make worked examples more effective:

- Worked examples should contain only partial solutions but the final answers are given. Students are to complete part of the solution; this is called the *completion problem effect.*
- Textual information should be integrated into the diagrams, if used, to avoid the *split-attention effect* that requires students to process disparate sources of information.
- Under the *redundancy effect,* getting students to process redundant materials has negative consequences on learning because of an increase in extraneous cognitive load. This suggests that worked examples should not include these materials.
- When variations are to be included, use formats that reduce extraneous cognitive load but increase germane cognitive load so that students are better able to distinguish between crucial features of the tasks from superficial ones (cf. Dienes' principles, see Chapter 2).
- Use the audio/visual rather than the visual/visual condition, which is called the *modality effect.* This applies only if the worked examples are to be used in multimedia presentation. In this case, "view animation + listen to narration" is more effective than "view animation + listen to narration + read online text" because the former combination avoids the split-attention effect. Mayer and

Moreno (2003) also noted that better transfer in learning occurs under multimedia instruction when words are presented as narration rather narration and on-screen text.

6.4 *Problem solving set*

The *problem solving set* or the *Einstellung effect* was first identified by Luchins in 1942: students continue to use the same rule in a mechanical way to solve problems even though more appropriate methods exist. This may arise for two reasons. First, these students believe that they have to use the rule that has been taught more recently. Second, they have been drilled on only one method through heavy use of worked examples with limited variability. In this case, the teacher does not encourage the use of alternative methods. This predisposition is evident in the two cases below.

Case 1. Having worked on several examples of finding the mean of numbers such as {28, 32, 35} by adding the numbers and dividing the sum by the number of items, the students are asked to find the mean of {48, 48, 48, 48}. They continue to solve it with the add-and-divide method, not realising that the mean is 48 by thinking about the meaning of *mean.*

Case 2. The students practise the quadratic formula on many questions and then continue to use it to solve equations such as $x^2 - 15 = 0$. They do not notice the simpler method of transferring the number to the right-hand side and taking square root.

Effective teachers must be aware of this tendency and counter it by getting students to compare and contrast different methods of solving the same problem. They should also be less rigid in marking students' solutions which may differ from the taught methods. By being more flexible in teaching and assessment, the teacher can promote creative and critical thinking among their students. In turn, they will enjoy the opportunity to share their different methods, to become more engaged in the lessons, and to feel less anxious about having to rigidly adhere to the given rules. This is also a simple case of how DI can be combined with student-centred learning.

7 Deliberate Practice

Deliberate practice should follow after explanation of worked examples. It helps students to master skills with accuracy, speed, and consistency in performance.

First, students must achieve some degree of automaticity in skills. This includes the ability to recall multiplication tables or expansion of trigonometric formulae by heart, so that they do not need to derive them from first principles when they need to use the results in subsequent steps of problem solving. By doing so, they can devote much of the limited working memory to process new information about the problem instead of *wasting* it on routine steps and of breaking off from their train of thinking. This also counters the *look-it-up* mindset, which increases extraneous cognitive load during problem solving.

Repetitive practice is often derided as *rote* and meaningless in contrast to constructivist approaches such as negotiation of meanings and group discussion. However, researchers on CHC distinguish repetitive learning from rote learning (e.g., Biggs, 1994; Watkins, 2007). Repetitive learning is practised by students in CHC to ensure accurate recall through repetition and understanding, whereas rote learning is to memorise something through mechanical repetitions. This distinction is important in understanding the place of repetitive practice in DI.

Students should also practise memorisation strategies. These include

- writing out definitions, formulae, and solutions over and over again to get them right,
- strengthening memory with flash cards,
- using mnemonics about mathematics rules, e.g., TOACAHSOH for the definitions of trigonometric ratios,
- making the rules memorable through songs and surprising connections, and
- elaborating links through the use of concept maps.

Theories about neuroplasticity and neurogenesis (e.g., Doidge, 2007) and neurological studies have shown that repetitions of skills can forge new neural pathways through growing new neurons and strengthening old ones, resulting in improved learning outcomes. Arrowsmith-Young (2012) reported many cases of how people have used brain-based cognitive exercises to overcome different types of learning disabilities, including mathematics. Dweck (2006) showed that teaching students about neural growth can help them develop a growth mindset, and this new mindset helps students to perform better in academic work than those who hold a fixed mindset.

Three types of deliberate practice are frequently used. They are enunciated below.

7.1 *Check seatwork or classwork*

Seatwork refers to students doing exercises during the lessons. The typical steps consist of the following:

a) Give clear verbal or written instructions on what to do.
b) Walk around *systematically* to look at how the students are tackling the problems. Try to cover as many students as possible in each lesson and do not ignore students siting at the corners or the back of the classroom.
c) Tick off the correct solutions in the students' workbook or worksheet to save marking time later on.
d) Correct mistakes promptly. If several students make the same mistake, call the class to attention and explain how to avoid the mistake.
e) If a mistake is unexpected, ask the student who makes it for an explanation. This will help the teacher know more about how students think.
f) When the students are practising several questions, use *progressive feedback:* after a few minutes, put the answer to question 1 on the board; walk around; after a few more minutes, put the answer to question 2 on the board, and so on. If necessary, call the class to attention to explain the difficult parts.

7.2 *Students work on the board*

The seatwork routine may include calling on several students to show their solutions on the board. The standard procedure is as follows:

a) Allow students a few minutes to work on the problem. Walk around to make sure that all the students have started on it.
b) When about half the class has attempted the problem, ask one or two students to put their solutions on the board. Select students who have obtained the right answer in order to focus on correct learning rather than mistakes and to avoid embarrassing those who are not successful in their attempts.
c) Avoid teaching only those who are putting up their work on the board or just stand in front of the class watching them doing so. Continue to walk around to check that the other students are still working on the problems.
d) When those at the board have finished writing down their solutions, call the class to attention and ask the students to explain their working or to go over the solutions. If the students are not used to speaking confidently and clearly in front of the class, training on this skill should be given. Make sure that all the students are paying attention to the explanation rather than continuing to complete their own solutions.
e) If the solution is correct and if the other students have little difficulty with the problem (this can be picked up while checking seatwork), do not waste time going over it in great detail.
f) Ask for alternative solutions to enhance understanding and creative thinking.
g) Deal with mistakes in an encouraging manner. Stress that it is important to make the effort to learn from mistakes rather than to avoid making any mistake at all cost.

7.3 *Homework*

Students are expected to spend some time outside class to complete assigned homework. Kaur (2011) found that Grade 8 students in Singapore believed that homework can serve six functions: improve understanding, practising and revising skills, improve problem solving, preparing for test, learning from mistakes, and extending mathematical knowledge. Cooper (2008) noted the positive relationship between the amount of homework students do and achievement. Rohrer (2009) found that this effect varies with the ways of assigning homework. My recommendation is to assign the same set of practice problems into the three categories below, in the ratio of 3:2:1. This gives more practice at the initial stage of learning a new skill to ensure overlearning.

- *Massed* (or *blocked*) practice of similar problems, immediately after initial learning of a new skill. At this overlearning stage, students are likely to make all kinds of mistakes, and this requires timely and detailed feedback from the teacher.
- *Distributed* (or *spaced*) practice of graded problems of the same skill spread over several weeks. This helps to maintain the skill and provides opportunity for students to rectify their mistakes through repeated practice.
- *Miscellaneous* (or *mixed*) practice using problems from different skills. This helps students develop the ability to discriminate between problem types and skills and to avoid the Einstellung effect.

Rohrer (2009) reviewed studies about the *spacing* effect and the *mixed* effect and concluded that both effects can increase the efficiency and effectiveness of practice and review, although the review covered only very few studies about mathematics. This overlearning-consolidation-discrimination approach to assign mathematics homework problems deserves further research.

Teachers face other challenges about homework. These include: late or no submission, poor quality work, complaints from parents about the perceived heavy homework load, considerable time spent on marking and giving feedback to homework, and record keeping when homework grades are included in summative assessment. Wong (1998b) described a workshop at which heads of mathematics departments of Bruneian secondary schools used de Bono's Thinking Hats to generate ideas to handle these homework challenges. Using Green Hat Thinking, they suggested letting students choose their own homework problems, grading for answers and effort, and conducting workshops for parents on homework in order to gain their cooperation. Automatising marking using ICT is another feasible way to reduce this chore; see Chapter 6. These suggestions can be converted to action plans and their effectiveness evaluated by the teachers themselves.

8 Address Student Mistakes

Making mistakes is inevitable at the initial stage of skill acquisition. Many student mistakes are due to underlying misconceptions (e.g., multiplication always results in bigger answers), flawed procedures (partially learned or poorly recalled), and applications of the wrong procedures to the particular problems. Unfortunately, some teachers attribute student mistakes to carelessness or slips, and they do not probe deeper to find the real causes.

A vast literature exists on how to deal with student mistakes in many mathematics topics at different levels. The methods are based on theoretical perspectives ranging from behaviourism to different versions of constructivism. This includes: Ashlock (2010), Barnard (2005), Borasi (1994), Hansen (2011), Olivier (1989), Ryan and Williams (2007, 2010), and Wong (2003a).

Students have different reasons for making the same mistake. Hence, it is important to give different explanations to deal with the mistake because the students may understand one explanation better than another. This is illustrated with the following five different explanations to help students deal with the classic algebraic error, $(x + y)^2 = x^2 + y^2$. This ubiquitous mistake is very resistant to remediation because its visual feature compels students to think of the distributive law: multiply inside the brackets. Since most of the students who make this mistake are weak in algebra, it is advisable to begin the remediation with a numerical example and then to generalise it to the algebraic case. For discussion purpose, the numerical example to be used is:

$$(x + 3)^2 = x^2 + 3^2 = x^2 + 9$$

Explanation 1. The standard explanation is to expand the perfect square, i.e., $(x + 3)^2 = (x + 3) \times (x + 3)$ [stress the meaning] $= x^2 + 6x + 9$ [expand]. This is the usual method used in teaching this expansion, and the mistake suggests that it has not been effective.

Explanation 2. Get students to submit different values of x and show that the expressions on both sides are not the same. This helps the students see that a mistake has been made but it does not show the correct expansion.

The two explanations above make use of the symbolic mode and the number mode. They are abstract and not easily internalised by the weak students. Explanation 3 includes the concrete and pictorial mode. Information processing theories suggest that when more modes are used to process information, the chances of noticing mistakes are higher.

Explanation 3. Apply the CPA or multi-modal strategy approach and show the expansion using concrete algebra tiles; see Figure 3.4(a). Focus students' attention on the fact that $(x + 3)^2$ is represented by all the pieces (16), whereas $x^2 + 9$ is represented by the shaded pieces only (10). This adds a visual feature to the schema of algebraic expansions.

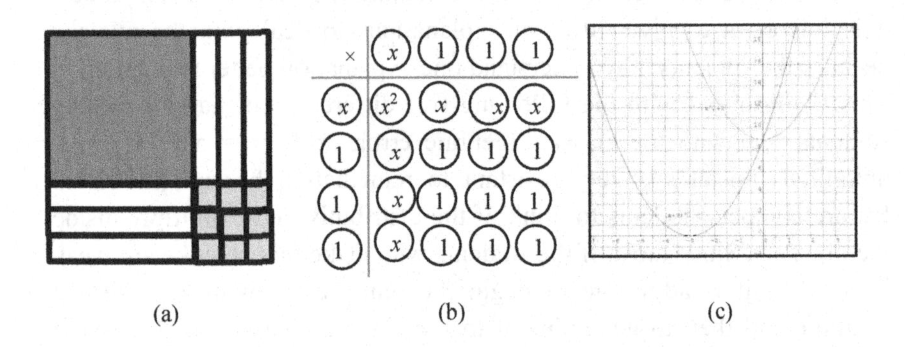

Figure 3.4. Expansion of $(x + 3)^2$: (a) algebra tiles (b) algebra discs (c) graphs

Explanation 4. Show the expansion using algebra discs (Leong et al., 2010); see Figure 3.4(b). Each algebra disc consists of a positive value on one side and its negative value on the inverse side. Unlike algebra tiles (Explanation 3), this tool does not make use of the concept of area; it is like completing an array by multiplying the values in the corresponding row and column.

Explanation 5. Plot the graphs of both expressions on the same scale; see Figure 3.4(c). Compare similarities and differences between these two graphs. Encourage the students to visualise the graphs in their mind.

After each explanation has been given, ask students to use it to expand similar expansions to consolidate mastery. Another effective learning strategy is to ask students to keep a journal of these explanations and to use them to complete homework and to review for tests.

Some teachers purposely write down wrong solutions on the board and ask the students to identify the errors. This is consistent with diagnostic teaching (e.g., Bell, 1993), which has been found to be beneficial. However, this should be used with extreme caution, because inattentive students may miss the teacher's explanation and think that what they have copied from board is correct.

As pointed out by Kuhn and Phelps (cited in Gredler, 2005, p. 263), "[c]onflicts and contradictions are encountered and resolved, not once, but many times over." Thus, teachers may have to correct the same mistake with different explanations over and over again until the students finally overcome those mistakes at their own time.

9 Skill Questions

To promote mastery of skills, teachers can ask *skill questions* in lessons or set them as prompts for students to write reflective review as they develop the skills over time. These questions should cover direct instruction, automatising standard skills, conditions for applying rules, worked examples, deliberate practice, problem solving set, student mistakes, and alternative explanations. Samples of skill questions include:

- What is the formula for …? What rule can you use? How do you know which formula to use?
- What is the first step …? Where do you begin?
- What is the next step …? What would you do next?
- What is the condition for using …?
- Can you think of another/simpler/more efficient method?
- Have you made this mistake before? What is the correction?
- Can you silently verbalise the steps for this rule?

Students should also be trained to ask themselves similar questions in order to monitor how well they have mastered the skills. This helps to develop metacognition.

10 Concluding Remarks

Students need to master standard skills so that they can use them with accuracy, speed, and consistency. This will reduce cognitive load on their working memory so that this limited resource can be used to process non-routine materials in the given problems. To do so, students must make sense of the concepts underpinning the rules, even though poor sense-making does not necessarily hinder mastery acquisition, as skills can be learned by repetitive and deliberate practice. Teachers can expedite skill mastery by using well-designed worked examples and exploiting a judicious combination of massed, distributed, and miscellaneous practice for homework and revision. These are some of the desirable characteristics of quality direct instruction.

Standard skills, however, should not be all there is in mathematics learning. In the next chapter, teaching of mathematical *processes* will be shown to build on strong conceptual foundation and consistent skill mastery.

Chapter 4

Processes: Sharpen Mathematical Reasoning and Heuristic Use

Mathematical processes are competencies higher than standard skills. The Singapore mathematics curriculum lists seven types of mathematical processes, but this chapter will examine only three types, namely, intuitive-experimental justification, deductive proofs, and heuristics, including model drawing. Question-and-answer and discussion can help students become versatile in mathematical reasoning.

If our reasoning leads us to the untrue conclusion, the fault lies with our reasoning. Walter Warwick Sawyer (1911-2008)

1 Mathematical Processes: Domain-Generic vs. Domain-Specific

Mathematical processes are competencies higher than standard skills. These higher order competencies are necessary if students wish to solve problems that are non-routine, unfamiliar, extended, or contextualised rather than routine ones and to communicate their solutions clearly to others. In short, these are the crucial competencies for good problem solvers and communicators.

At the time when the Singapore Pentagon framework was being designed in 1989, there was considerable debate about whether domain-generic thinking programmes such as de Bono's lateral thinking (de Bono, 1992) or domain-specific school subjects are the optimal way to help students become better thinkers or problem solvers. Advocates of the generic approach believe that thinking and problem solving skills are general competencies, and they can be learned per se and applied to different school subjects or daily life (recall the saber-tooth curriculum discussed in Chapter 1). They also think that the acquisition of content knowledge is not sufficient to inculcate these generic skills. The purported advantage of this approach is that these generic competencies will better prepare students for the unpredictable, rapidly changing world in the future. The recent focus all over the world on so-called *21st century competencies*[13] and the infusion of desirable social and even spiritual values in all school subjects including mathematics (e.g., Winter, 2001; Wong, 2003b) are the latest examples of this approach.

Advocates of the domain-specific approach, on the other hand, argue that thinking skills with the same labels actually involve different competencies in different school subjects. For instance, critical thinking is different in mathematics or in history: it is about making inferences using logic in the former and using empirical data in the latter. A strong advocate of the domain-specific approach is Sweller, whose view was cited in a recent review of the Australian curriculum (Donnelly & Wiltshire, 2014) as follows:

> There is little more useless than attempting to teach generic thinking skills and expecting students to be better thinkers or problem solvers as a result. Despite decades of work, there is no body of evidence supporting the teaching of thinking or other generic skills. (p. 125)

Back in late 1980, the Singapore mathematics curriculum committee addressed a similar issue within mathematics itself, namely, whether generic mathematical competency such as proving and heuristics should be taught on their own or embedded into the different content topics. After much deliberation, the committee decided to highlight this distinction by using the two terms, *skills* and *processes,* to cover the specific and generic aspects of mathematical competency respectively. Teachers are expected to repeatedly infuse these processes into lessons of all the topics. The repetitions provide the learning experiences to help students develop these processes under different guises.

---

[13] See the Singapore approach at http://www.moe.gov.sg/education/21cc/

The specific competencies covered under this *Processes* factor in the Singapore mathematics curriculum framework have changed slightly over the past two decades, but the key ones remain the same. They are: *reasoning, communication, building connections, thinking skills, heuristics, applications,* and *modelling*. Curricula in other countries also cover similar processes under different labels, such as *mathematical practices* (US), *working mathematically* (Australia, England), *reasoning mathematically* (England), *habits of mind,* and *mathematics competencies* (Denmark).

This chapter will deal with mathematical reasoning and heuristics, and Chapter 5 will cover application and modelling.

2 Mathematical Reasoning: Definition

In Singapore, reasoning has been suggested as the fourth *R* in the curriculum to "nourish[es] children's natural sense of intellectual curiosity." [14] Mathematical reasoning is "the ability to analyse mathematical situations and construct logical arguments" (Ministry of Education, 2012, p. 15). The NCTM (2000) gave a more elaborate description that combines reasoning and analytic thinking:

> People who reason and think analytically tend to note patterns, structure, or regularities in both real-world situations and symbolic objects; they ask if those patterns are accidental or if they occur for a reason; and they conjecture and prove. (p. 56)

The key ideas are patterns, conjectures, logic, proofs, and justifications. The main function of reasoning is to justify to oneself and others why certain statements are true (valid) or false. This corresponds to *relational understanding,* defined by Skemp (1976) as "know both what to do and why" against *instrumental understanding,* which is about "rules without reasons" (p. 21). He mentioned that relational understanding has at least four advantages. One of the advantages is that teaching mathematics for relational understanding can help students

---

[14] http://www.moe.gov.sg/media/speeches/2006/sp20060417.htm

develop the confidence and habit of mind that, should they forget a rule, they will try to work it out from first principles.

For teachers, this habit of reasoning things out for oneself should be fostered during pre-service training. The episode below happened during practicum and shows that the pre-service teacher had not developed this habit. Mary (not her true name) held a BSc degree in Mathematics. She had just taught the Sine Rule using only worked examples but no justification or proof.

I: Why didn't you show the proof?

Mary: I don't know the proof. My school teacher did not teach it to us.

I: But you are a maths major. Can you prove it yourself?

Mary: Don't know.

I: Please prove it yourself or look it up in the textbook. Explain it to the class in the next lesson.

Some teachers do not teach mathematics for relational understanding because they think that doing so takes too much curriculum time from the need to *cover the syllabus*. Other teachers lack content knowledge, and teachers like Mary have not internalised the belief that mathematics is essentially about reasoning. These three groups of teachers need to receive different professional development to address their shortcomings.

Rigorous mathematical reasoning is based on deduction. However, psychological research has shown that most students do not acquire formal deductive reasoning until they are at the upper secondary level, if at all (e.g., van Hieles, 1986). Nevertheless, students must be exposed to the rudiments of reasoning from primary level onward because it takes time and much practice to build the reasoning habit of mind. At the primary level, reasoning is to be taught through an intuitive-experimental approach (*inductive thinking, experiential justification*), even though it is mathematically not rigorous.

3 Intuitive-Experimental Justification

Inductive thinking refers to making a generalisation and justifying it by examining several cases that satisfy the targeted properties. This type of argument is based on finding regularities or patterns in the data or cases given. It is a very powerful reasoning strategy widely used to help students make sense of new concepts and rules and to notice regularities in their daily experiences. Indeed, mathematics is often considered the science of patterns (Devlin, 1997).

This pattern-seeking ability is evident even in very young children when they arrange concrete materials according to shape, colour, or other attributes. They can be asked to describe what they see, to repeat the patterns, and to extend them. At a later stage, they work with patterns given in figures (e.g., ♦ ■ ★ ♦ ■ …), numbers (e.g., 1, 3, 5, 7, …), or symbols (e.g., x, x^2, x^3, …). These experiences prepare them to identify more complex patterns later on.

Since only a few cases are to be used in these pattern-seeking and pattern-extending activities, a pedagogical issue concerns how many cases are *enough* to infer the valid conclusions. For students who begin to learn the logic of intuitive justification, four to six cases are usually sufficient if they show clear regularities. Students who give only one or two confirming cases should be challenged to come up with more cases. Ask them to decide for themselves how many cases will convince them of the validity of a claim, and compare their suggestions. This will generate fruitful discussion about justification, persuasion, and proof. It also helps them to focus on their own sense-making and learning strategy, in a way that aligns with the constructivist epistemology and metacognition.

3.1 *Teaching inductive-experimental justification*

A combination of direct instruction and guided discovery is more effective than either strategy used on its own. A suggested sequence is given below.

a) Direct instruction: Briefly explain what is to be expected, not necessarily the final result. This counters the tendency that, under unguided discovery, students observe only the obvious or irrelevant features and miss the mathematical points. I call this *not seeing what is not expected* or the reverse, *to know is to see.*

b) Give students different examples to work on, in order to increase variability about mathematical or perceptual features (Dienes, 1964). They can complete the activity individually or in groups. Non-examples should also be assigned.
c) Call the class together to discuss how the different examples and non-examples can be used to justify the intended result. If measurement is involved as is common in activities about mensuration and geometry, discuss the effects of measurement errors.
d) Once the justification has been understood, provide further work using online materials, which are more accurate than physical ones.
e) Provide practice to consolidate mastery of the new skill or concept.

3.2 *Examples of inductive-experimental justification*

This section looks into four examples of inductive-experimental justification.

Example 1. Sum of even and odd numbers
To justify that the sum of an even number and an odd number is odd, primary school students can make up a table, fill it with even and odd numbers and their sums, and notice that the sums are always odd. This inductive process can be enhanced by using blocks: there is always an odd one left after the blocks are arranged in pairs.

Example 2. Addition and subtraction of integers
This is a very difficult topic for teachers to teach and students to learn. The cognitive difficulty is due to the abstract nature of negative numbers. Historically, famous mathematicians up to the 18th century thought that negative numbers were greater than infinity (Heeffer, undated). Similar

conceptual confusions may be dormant in students' mind, but they can cause cognitive dissonance once activated through various rules.

The pedagogy difficulty arises because of poor teaching moves that confuse the students. Nine of these poor teaching moves are discussed below; there may be more. When more than one poor move is used in a lesson, the confusion to students also increases drastically.

a) The stories or models used to introduce integer operations are contrived or difficult to understand; for example, *to subtract a negative number is like bad people leaving a place, so it is like adding good people, converting subtraction to addition*. This leads to confusion of meanings about the operations.

b) Complicated statements are used to describe the rules, and students cannot understand these statements when they do not have the necessary concrete experiences to support sense-making; for example, *to add two numbers of different signs, find the difference between their positive values and give the sign of the numerically larger integer*. The concept of *absolute value* is sometimes used, and this presents additional hurdle to some students.

c) Rules are given in algebraic form when the students have not yet learned much about algebraic manipulations; for examples, $a + (-b) = a - b$, if $a > b$, and $a + (-b) = -(b - a)$, if $b > a$.

d) Notations and terms about integers are used inconsistently; for example, *minus* is used to mean both an operation (subtraction) and a property (negative), e.g., -3-4 versus -3(-4); both expressions seem to trigger off the "minus times minus equals plus" rule.

e) Addition and subtraction are switched without due attention to meanings; for examples, 4 + (-3) is treated as 4 – 3; this does not well work for 3 + (-4) = 3 – 4, because at this stage, the students do not know the meaning or the rule of subtracting a number from a smaller one.

f) Multiplication rules are used to justify the change of signs; for example, – (-3) is + 3 in the expression 4 – (-3) = 4 + 3, because of the same "minus times minus equals plus" rule. Although this rule works well in similar examples, it does not justify why the rule could be applied from multiplication to subtraction. The confounding of several operations within the same expression is a particularly poor teaching move.

g) There are weak attempts to link the operations on negative integers to the same operations on positive integers so that meanings can be built on prior knowledge; for example, subtraction is about *taking away* at primary levels, but this concept is not used consistently for subtraction of negative integers.
h) Teachers do not explain enough numerical examples to cover all possible cases; in fact, the more difficult ones are left, perhaps inadvertently, as exercises for students to work out on their own. See the example given in (e) above. This lack of coherent coverage might work for high ability students but is disastrous for most students; see Chapter 2, Section 2 about a similar approach for concepts.
i) Teachers rush through the initial lessons on integers and do not give sufficient revision to consolidate the skills and concepts.

Teachers should be aware of these weaknesses and plan lessons to avoid them. Integer operations are used again in algebraic expansions and other topics, so poor mastery of this topic can result in subsequent learning difficulty. Teachers may be compelled to re-teach this topic for remediation; this additional time and effort can be averted if more care and time are given to teach this topic at the beginning.

Three teaching approaches can be used: *set* (how many?), *number line* (how large?), and *number patterns* (e.g., Long, DeTemple, & Millman, 2012). Each approach has its pros and cons, and research has not identified the *best* approach, so it is up to the teachers to learn these methods and use them to deal with the learning difficulty encountered by their students.

Let us first consider number pattern because of the power of patterns and ease of use. The basic steps are as follows.

i) Students *write* down the sums: $3 + 2 = \square$, $3 + 1 = \square$, $3 + 0 = \square$, $3 + (-1) = \square$, etc. This extends their prior knowledge about addition of positive integers to negative integers; this addresses move (g) above. Furthermore, the meaning of addition is preserved, without bringing in subtraction, to avoid move (e).

ii) Students describe the pattern *in their own words;* for example, as the second addend decreases by 1, the sum also decreases by 1. Instead of giving students complicated statements to remember, which is move (b), let students state the rule in ways that make sense to them. When their statements are shared in the class, students learn to better communicate their ideas.

iii) Repeat the above with negative integers as the first addend. This addresses move (h).

iv) Ask the students to revise their rules after they have covered all possible combinations. This also addresses move (h).

v) Add the *diagram* mode by creating an online slider to show the pattern. This is helpful to visual learners.

vi) Practise the new skills with fractions and real numbers. This ensures that the work is not limited to only integers.

Table 4.1 can be found in worksheets using this pattern approach. Students begin with one cell, say, 3 + 2 and work to the left by adding smaller and smaller numbers and noticing the pattern.

Table 4.1

Matrix for addition of integers

				second	**addend**	
	+	**-2**	**-1**	**0**	**1**	**2**
	3			←	←	
first	**2**					
addend	**1**					
	0					
	-1					
	-2					

This format is not effective because many students mechanically fill in the numbers in the cells, *without paying much attention to the actual sum involved in each cell.* Instead, ask them to write down the individual sums, as suggested in step (i) above; doing so incurs additional information processing which helps them better notice the patterns. This makes use of the psychological process called *generation effect* (McNamara, 1995). According to this theory,

> having students generate to-be-learned information themselves rather than simply copying or reading the information enhances both short-term ...and long-term ... retention of information in various situations. (p. 307)

Next, consider the teaching of these two operations by moving numbers along a number line. The details can be found in many textbooks, but they usually treat the motions in isolation and do not help students see the connections among them. One way to link these motions is shown in Figure 4.1. This should be arrived at through whole class discussion and not given to them as another rule. For example, the figure shows that adding a positive integer is equivalent to subtracting its negative equivalent.

Operation	**Second addend**	
	Positive	Negative
Addition	Move to right →	***Move to left ←***
Subtraction	***Move to left ←***	Move to right →

Figure 4.1. Links between addition and subtraction of integers on a number line

The third method is to use positive and negative cards and the neutralisation principle: a combination of +1 and -1 has zero value. The Singapore Ministry of Education has created the manipulative called AlgeDisc™ and an online version[15] to facilitate this approach. The latest Singapore textbooks all use this approach. Leong and colleagues (2014) named their models the AlgeCards. Both versions have +1 and -1 in different colours on opposite sides of the disc or card. The main difference is that the disc is round and the card is square; the card version also embeds the area concept. An engaging, simple, and cheap way is to

---

[15] http://algetools.moe.edu.sg/

ask students to create their own cards instead of purchasing commercial ones.

Finally, the multi-modal strategy thinkboard can be used to summarise what is taught after several lessons. A completed thinkboard is shown in Figure 4.2. Note that the *story* is to be completed at the end of justifying the rules, and *not* at the beginning as a model for addition; this avoids the poor teaching move (a) described above.

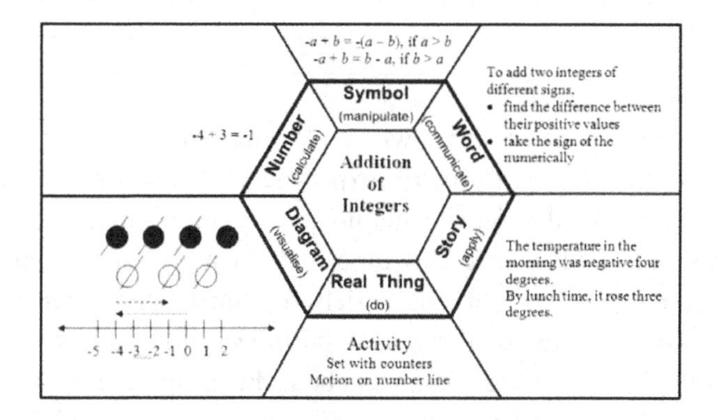

Figure 4.2. Multi-modal strategy thinkboard: addition of integers

Example 3. Sum of the interior angles of an n-polygon

First, justify that the sum of (interior) angles of a triangle is 180°. This is typically done by asking students to draw different types of triangles (equilateral, isosceles, acute-angled, right-angled, or obtuse-angled), measure the angles, and show that the sum is close to 180°. This follows the *measure-calculate-discuss-conclude* routine. Another popular justification move is to cut out the three angles and assemble them along a straight line. Both activities can be readily completed by primary school students.

Next, this property about triangle is used to justify the rule for an n-gon. Begin with quadrilaterals. Students use the same measure or assemble method to conclude that the sum of interior angles of a quadrilateral is about 360°, which is 2 × 180°. Repeat this for other polygons. Tabulate the results as in Table 4.2; complete only the first three columns at this stage. It is quite easy to arrive at the formula $(n - 2) \times 180°$ through this inductive approach.

Table 4.2
Sum of interior angles of n-gon

Polygon	No. of sides, n	Sum of interior angles	No. of triangles
Triangle	3	180°	1
Quadrilateral	4	× 180°	2
Pentagon		× 180°	
Hexagon			
n-gon			

Example 4. For a circle, $C = \pi d$, where C = circumference, d = diameter The pedagogical idea is to justify experimentally that C/d is a constant for all kinds of round objects. This invariance allows mathematicians to *define* this ratio with a special symbol, which was first written as π by William Jones in 1706 and was widely accepted after Euler used it in 1748[16]; in Greek, π sounds like *p* as in *periphery* or *perimeter*.

Almost all the teaching resources about this property require students to measure C and d of various round objects, compute C/d for these values, and conclude that these computed values are approximately 3. To save class time, students work on different round objects, and the results pooled together to show that the size of the circle does not matter. This inductive approach can be strengthened by adding an extra step: compute the values of other expressions, such as $C + d$, $C - d$, and Cd. By comparing these four sets of expressions, one can notice the invariance of C/d much better. Since the other three expressions do not give similar values for different circles, it is not helpful to give them special symbols as for C/d. This *compare-and-contrast* thinking skill is very helpful here, and it should be used frequently in teaching other topics.

The following exercise will be of interest to the more able students. Inscribe an equilateral triangle within a circle of diameter d. Let P be the perimeter of the triangle. Measure d and P. Produce a table of values of P/d for equilateral triangles constructed inside different circles. Does it produce a stable value? Repeat for squares and other regular polygons.

---

[16] http://en.wikipedia.org/wiki/Pi

This provides an empirical introduction to the idea that a circle can be approximated with a series of regular polygons; see Wong (1997a) for details. Ancient mathematicians like Archimedes used similar limiting processes to *prove* that C/d is a constant for different circles. Hence, it is meaningful to assign to this constant a special symbol, as noted above.

The inductive approach provides a powerful avenue for students, especially kinaesthetic learners, to make sense of abstract mathematics by working with patterns. They will begin to appreciate the functions of justification in mathematical thinking. However, they must learn that inductive thinking is incomplete, which will be discussed in the next section, and this realisation leads to the need for rigorous proofs in Section 4.

3.3 *Caveats about inductive thinking*

As noted above, inductive thinking covers only a few cases, so an obvious concern is whether statements about observed patterns apply to *all* relevant cases. That highly convincing pattern may fail in the next case is illustrated by the well-known black swan phenomenon: to meet a black swan after seeing only white swans on numerous occasions. In mathematics, too, obvious patterns do fail, and students need to learn that pattern does not constitute proof. This idea may not have been given the attention it deserves in mathematics teaching.

First, consider the following *prime number formula*. The question is whether the expression $n^2 + n + 41$ always generates a prime number.

- After substituting $n = 0, 1, 2$, to get 41, 43, 47, and finding that they are all primes, some students jump to the conclusion that the expression does generate only primes.
- Ask them to ask more cases. Indeed, the next 38 values are primes, but many students stop after checking only a few more values, and this seems to strengthen their conviction.
- At this stage, ask students to find a value of n so that it is *not* a prime. The perceptive students will notice that when $n = 41$, the expression gives a composite number because it has 41 as a factor.

- Finally, discuss what this activity says about inductive thinking and the need for deductive argument.

The following four cases look into different aspects of the above issue. Cases 1 and 2 show that inductive thinking can lead to the correct results, provided that the underlying structure is known or can be established. Case 3 shows how an extension to an obvious number sequence can fail when it is divorced from the structure of the given problem context. Finally, Case 4 illustrates the flaw in the popular "guess the next number" puzzle used in textbooks and IQ tests.

Case 1. Figural pattern problem
Figural pattern problems are those that show the first three or four figures of a pattern and require students to demonstrate their generalisation skills by writing down the rule using algebra. These figural problems are very popular in textbooks, research studies, and public examinations (e.g., Chua & Hoyles, 2014; Leong et al., 2014). The figural problem shown in Figure 4.3 is a particularly popular one. Use three toothpicks to form the first triangle. Add more toothpicks to make the next two figures. How many toothpicks are there in the *n*th figure?

Figure 4.3. Number of toothpicks for the figural problem

The common approach is to build a table of values (see Table 4.3) by counting the number of toothpicks for the first few figures and note the pattern, which is quite obvious here.

Table 4.3
Number of toothpicks for the figural problem

n	**1**	**2**	**3**	**4**
T(*n*)	3	5	7	9
Pattern	+ 2	+ 2	+ 2	+ 2

To obtain the rule, students usually manipulate the values until they get the rule, $T(n) = 2n + 1$. To carry out the *look back* stage of Polya's problem solving model, students may check that the rule gives the correct answer for the next few values of n. This is the end of the exercise for most students, as expected by the teachers. Their learning experience is to obtain an algebraic rule based on the set of numbers, without referring to the initial problem context.

The next step, rarely done, is to help students understand the idea that patterns can lead to the correct answers if there are sound explanations of the underlying structure. Ask them to explain the rule by referring to the way they counted the number of toothpicks at each step. One way to count is to begin with one toothpick (1) and add two toothpicks at each stage (giving $2n$), resulting in $1 + 2n$; see Figure 4.4(a).

Figure 4.4. Two ways of counting the number of toothpicks for the figural problem

Two other ways of counting are:

- Begin with 3 toothpicks; at each subsequent stage, add 2 toothpicks: this gives $3 + 2(n - 1) = 2n + 1$.
- Add three toothpicks at each stage, but subtract the $(n - 1)$ overlapping ones: this gives $3n - (n - 1) = 2n + 1$; see Figure 4.4(b).

These different ways of counting the toothpicks lead to the same rule. The way of counting defines the underlying structure of the problem and the rule.

Let students share their ways of counting in order to promote flexible and creative thinking, meaningfulness of algebraic manipulations, enjoyment, and appreciation of mathematical justification. It also makes the algebraic rule *visible* when students learn it through kinaesthetic actions on concrete manipulatives, as recommended in the CPA and multi-modal strategy.

Figure 4.4(b) was used in the TEDS-M study (Tatto et al., 2012). Only 33% of the Singapore sample (n = 380) and 30% of the international sample (n = 14,000) were able to relate the algebraic rule to the counting process (Wong, Boey, et al., 2012). This weak performance suggests that the verification step of pattern spotting should be given stronger emphasis.

An extension to the above activity is to change triangles to squares, pentagons, and n-gons. Encourage students to discover the general rule for the kth figure of an n-gon: $(n - 1)k + 1$.

Case 2. Chord problem

A chord cuts a circle into two regions. Two chords cut it into four regions and so on. What is the greatest number of regions made by n chords? Let $T(n)$ be the largest number of regions for n chords. The first few values are easily generated; see Table 4.4.

Table 4.4

Greatest number of regions given n *chords*

n	0	1	2	3	4
T(n)	1	2	4	7	11
Pattern	+ 1	+ 2	+ 3	+ 4	

On the basis of extending the number sequence given in the table, the pattern is to add 2, 3, 4, … at each stage, or mathematically, $T(n) = T(n - 1) + n$, for $n \geq 1$. This is a *recursive* rule.

The challenge is to explain this pattern in terms of how the regions are generated according to the given context. One could argue thus. A new region is created when a new chord is added to the circle or when it cuts an existing chord. To obtain the greatest number of regions, the new chord must intersect all the existing ones. This gives an additional $1 + (n - 1) = n$ regions. Hence, $T(n) = T(n - 1) + n$. Once again, there is a sound mathematical explanation for the pattern in the number sequence.

Deriving the explicit formula, $T(n) = 1 + \binom{n+1}{2}$, is an algebraic problem, and it is not directly related to the problem context.

Case 3. Dot problem

Take any two points on a circle. Join them with a chord. This divides the circle into two regions. Repeat this process with 3, 4, 5, ... points. In each case, find the greatest number of regions formed inside the circle. The first five values are shown in Table 4.5, where R(n) be the greatest number of regions for n points on the circle.

Table 4.5

Greatest number of regions given n *points on the circumference*

n	**1**	**2**	**3**	**4**	**5**
R(n)	1	2	4	8	16
Pattern	2^0	2^1	2^2	2^3	2^4

The number sequence suggests the *obvious* rule, $R(n) = 2^{(n-1)}$. Is there a sound mathematical explanation for this pattern based on the way the regions are constructed? Unfortunately, there is none, and in this case, the pattern breaks down when $n = 6$: instead of $2^{6-1} = 32$ regions, there are only 31 regions; see Figure 4.5. Indeed, many teachers and students find this *breakdown* quite incredible!

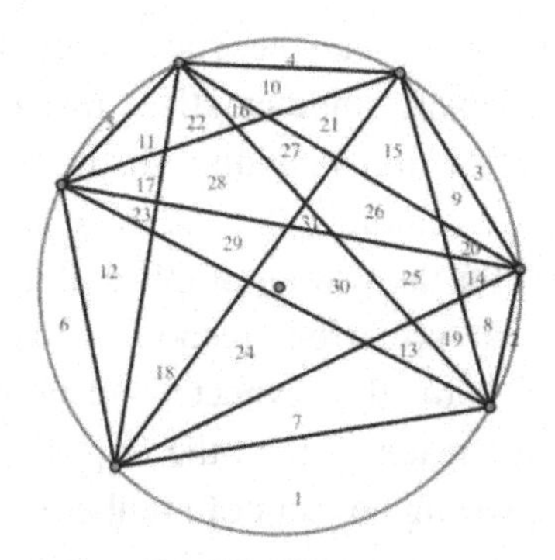

Figure 4.5. Largest number of regions with 6 points on the circumference

The correct formula is $R(n) = 1 + \binom{n}{2} + \binom{n}{4}$. See Kaplan and Kaplan (2007) or Scholtz (2011) for a mathematical explanation.

Case 4. Guess the next number puzzle
Consider the typical popular "guess the next number" puzzle. What is the next term of the sequence {1, 2, 4, 8, 16 ...}? Case 3 above shows that there is no unique answer: if the implicit and *obvious* assumption is powers of 2, then the answer is 32; if the assumption is the construction of the regions as in Case 3, the answer is 31. A less obvious rule is, $2^n + n(n-1)(n-2)(n-3)(n-4)$.

In general, given the first k terms of a sequence, one can continue the sequence in any way one likes in this way: write down any number for the $(k+1)$th next term and use the Lagrange Interpolating Polynomial[17] to produce a polynomial that passes through all the $(k+1)$ terms. This shows that mathematically, such puzzle has no unique solution. However, out of these infinite possible extensions, only a few are mathematically interesting or useful. One way to check this is to enter the first few terms into *The On-Line Encyclopedia of Integer Sequences*[18] and study the output.

Understanding about the nature of this puzzle was explored in a study conducted in Brisbane (Wong, 1984; 1987). The task was:

What is the next term of the sequence {7, 16, 25, 34, 43, 52, ...}?

Four options were given, and these are reproduced in Table 4.6. A sample of mathematicians, mathematics teachers, and pre-service teachers rated the level of understanding displayed in each option on a 0-4 scale (0 = No understanding; 4 = Good understanding). Table 4.6 shows the mean scores for the whole sample and individual groups.

The mathematicians preferred response (b) that spelt out the alternative assumptions to the expected response (a), whereas the teachers and pre-service teachers held the opposite view. Some of these teachers may have the wrong concept of number sequence; this is evident from this response: "sequences are concerned with integers and not with digits."

---

[17] http://mathworld.wolfram.com/LagrangeInterpolatingPolynomial.html
[18] http://oeis.org/

Table 4.6

Mean scores of understanding on a number sequence task

Responses	Whole (90)	MA (16)	MT (46)	FT (28)
a) This is an AP of common difference = 9. The next term is 61.	3.8	3.3	3.8	3.9
b) If it is an AP, the next term is 61. It can also be 59, because the sum of the digits of each term is divisible by 7.	3.5	3.6	3.4	3.6
c) It can be anything. You can't tell from six terms only. You must be given the rule.	0.9	2.6	0.7	0.8
d) This is a set. In set, you can arrange the terms in any way you like, say {7, 34, 52, 16, …}. Don't know the answer.	0.5	0.7	0.4	0.5

Note: MA: Mathematicians; MT: Mathematics Teachers; FT: Future Teachers.

The mathematicians rated the theoretically correct (c) as displaying better understanding than both groups of teachers who virtually rejected it as an acceptable response. One teacher wrote: "He/She is right but is failing to see the point of the exercise." This raises the question about what mathematical and pedagogical purposes are served by engaging students with such puzzles. Some teachers considered that six terms were sufficient to determine the rule, once again displaying misconception about number sequences. Option (d) was considered to display virtually no understanding of the task, although it is correct to note that the same symbol has been used to represent different mathematical concepts.

3.4 *A brief summary about inductive thinking*

It is important to help young students appreciate the power and applications of the intuitive-experimental approach using patterns, but mathematics is more than pattern spotting (Gardiner, 1987) or pattern extension. The above examples highlight a tension between mathematics rigour and pedagogical practice. The ubiquitous number sequence puzzle serves the pedagogical purpose to develop generalisation skill among students at elementary levels, but its use is mathematically incorrect. One way to resolve this tension is to guide students to progress from informal, intuitive thinking to mathematically rigorous practices. This naturally leads to the teaching of deductive proofs.

4 Deductive Proofs

Deductive reasoning (*deductive thinking, deductive logic, logical deduction*) is the opposite of inductive reasoning. It involves applying logical principles to premises in order to arrive at certain conclusions; in other words, from general premises to special cases. Deductive proof is a sequence of steps, each step following logically from the previous ones, resulting in the intended conclusions. An elegant proof also makes clear how powerful ideas are organised to tell a mathematical story or theme. Proof is, indeed, the heart of mathematical reasoning.

There are several standard proof techniques. They are: direct proof, proof by exhaustive enumeration, proof by contrapositive, proof by mathematical induction (different from inductive thinking), and proof by contradiction. However, most mathematics curricula around the world worldwide do not cover these proofs in a systematic way. Rudimentary proofs can be embedded in many topics from upper primary to secondary level. For examples, in her book, Waring (2000) provided detailed lesson plans and proof activities, organised under four phases: *convince a friend, convince a penfriend, towards formal proof,* and *proof for all.* These are good models for teachers to design similar proof activities for their students.

The next section shows how proof can be included as extensions into the first three intuitive-experimental activities discussed in Section 3.2.

4.1 *Some proof examples*

Example 1. Sum of even and odd numbers

Let a be an even number and b an odd number. Then, $a = 2m$ and $b = 2n + 1$, so their sum, $a + b = 2(m + n) + 1$, is odd. After this demonstration, let students discover the sums of different combinations of even and odd numbers and prove their conjectures.

Example 2. Addition and subtraction of integers
The inductive-experimental approach is used to extend the meanings of these two operations to cover both positive and negative integers. Under this approach, there is nothing to prove.

At tertiary level, however, the set of real numbers is treated as a *field* satisfying several axioms[19]. It is possible to use these axioms to prove results such as (-1) × (-1) = 1. Some mathematicians (e.g., Wu, 2011) argue that this axiomatic approach should also be used in schools. Although this has not been implemented in many countries, mathematics teachers should know about this approach and be able to explain it to the mathematically inclined students to stimulate their interest for rigour.

Example 3. Sum of interior angles of an n-polygon
First, prove that the sum of angles of a triangle is 180°. The standard proof uses an exterior angle and angles with parallel lines.

Next, apply this property to prove the rule for an n-gon. Insert the correct values into the last column of Table 4.2. An n-gon can be divided into $(n - 2)$ triangles and the sum of the interior angles of these triangles is $(n - 2) \times 180°$. As an illustration, the standard way to dissect a pentagon is shown in Figure 4.6(a). Figure 4.6(b) shows a dissection from a point inside the pentagon; ask students to give a proof using this dissection. This alternative aims to promote creative thinking.

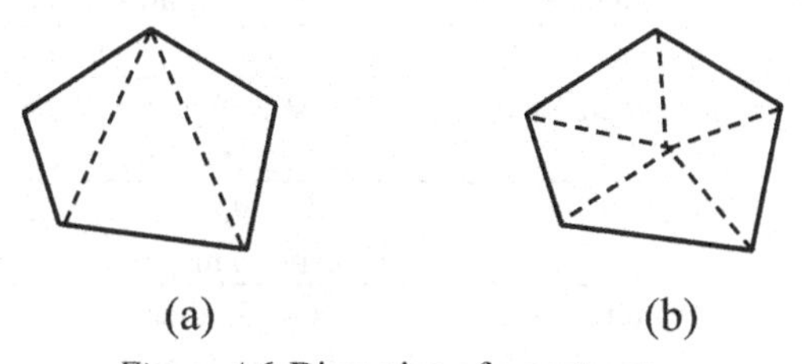

Figure 4.6. Dissection of a pentagon

[19] http://www.math.caltech.edu/~2011-12/1term/ma001a/11Ma1aNotesWk2.pdf

4.2 *Logical forms*

The canonical example of a deductive argument is based on Aristotle's list of syllogism. The most basic one is as follows:

a) All men are mortal. (*premise* 1)
b) Socrates is a man. (*premise* 2)
c) Therefore, Socrates is mortal. (*conclusion*)

Mathematics statements are often expressed in the *if-then* format: if (premise) then (conclusion). These are called *implications.* However, most statements used in school mathematics are not written in this format; instead, declarative versions are used. The following example shows the difference:

- (Implication) If m and n are negative integers, then mn is positive.
- (Declarative) The product of two negative integers is positive.

Students at upper secondary levels should learn the four basic logical forms, namely, *implication, contrapositive, converse,* and *inverse,* and their relationships. Table 4.7 gives examples of these four formats.

Table 4.7
Format of mathematics statements

Name	Format	Example (True/False)
Implication	If p then q.	If $a = 3$, then $a^2 = 9$. (*implication*, true)
Contrapositive	If *not* q then *not* p.	If $a^2 \neq 9$, then $a \neq 3$. (*contrapositive*, true, equivalent to *implication*)
Converse	If q then p.	If $a^2 = 9$, then $a = 3$. (*converse*, false, because a may be -3)
Inverse	If *not* p then *not* q.	If $a \neq 3$, then $a^2 \neq 9$. (*inverse*, false; *inverse* is equivalent to *converse*)

Mathematics teachers should understand these logical equivalences, but it seems many do not. In the *Mathematical Vitality* study (Wong, 1990b), about 43% and 55% of Singapore and Australian pre-service teachers respectively believed wrongly that the converse of a theorem is always a theorem that this may or may not need to be proved. A similar result was reported by Lim-Teo (2012) in a recent study of about 60 Bachelor degree mathematics students at Singapore National Institute of Education. These two studies suggest that the *implication-converse* confusion has remained problematic over the past 25 years in Singapore. Two plausible reasons for this weakness are that (a) the converses of many geometry theorems found in the secondary school curriculum are also true, and (b) these pre-service teachers may not have encountered sufficient cases of false converses to grasp the difference.

In the Singapore mathematics curriculum, the implication-converse idea occurs in several topics (see Table 4.8), but the term *converse* is not used; perhaps it should be. Teachers who have not understood the distinction may have taught the wrong mathematics to their students.

Table 4.8
Selected topics involving converse

Topic	Theorem	Converse
Parallel lines	Corresponding angles for two parallel lines are equal.	If the corresponding angles are equal, the two lines are parallel.
Pythagoras' Theorem	In triangle ABC, if C is a right angle, then $a^2 + b^2 = c^2$.	In triangle ABC, if $a^2 + b^2 = c^2$, then C is a right angle.
Angle in a semicircle	If C lies on the semicircle with diameter AB, then $\angle ACB$ is a right angle.	In triangle ABC, if C is a right angle, then C lies on the semicircle with AB as diameter.
Zero Product Rule	If $ab = 0$, then $a = 0$ or $b = 0$ (or both).	If $a = 0$ or $b = 0$, then $ab = 0$.

Two converse are discussed below to show how they may be taught.

4.3 *Converse of Pythagoras' Theorem*

Although the term *converse of Pythagoras' Theorem* does not appear in the curriculum, students are required to determine whether a triangle is right-angled or not, given the lengths of three sides. Some teachers think that it is the same as the Pythagoras' Theorem, while others do not know the proof. Indeed, the oft-cited story about ancient Egyptians making a right angle using a string 12 units long is about the converse and *not* the theorem itself! The following *inductive* → *deductive* sequence, as discussed in earlier sections, works for pre-service teachers.

a) Students choose their own set of two values, say $a = 4$ and $b = 7$, and compute c such that $c^2 = a^2 + b^2$.
b) They construct the triangle with sides a, b and c, using their values. Measure its three angles. Is there a right angle?
c) Hold a class discussion about their conclusions from (b) above. Highlight the converse property irrespective of the values chosen.
d) Explain the deductive proof, which is based on Euclid; refer to Figure 4.7.
   - Given ΔABC with $c^2 = a^2 + b^2$.
   - Construct ΔXYZ such that $XZ = b$ and $Z = 90°$ and $YZ = a$.
   - Apply Pythagoras' Theorem to ΔXYZ: $XY = c$.
   - Hence, ΔABC and ΔXYZ are congruent (SSS).
   - Therefore, $C = Z = 90°$.

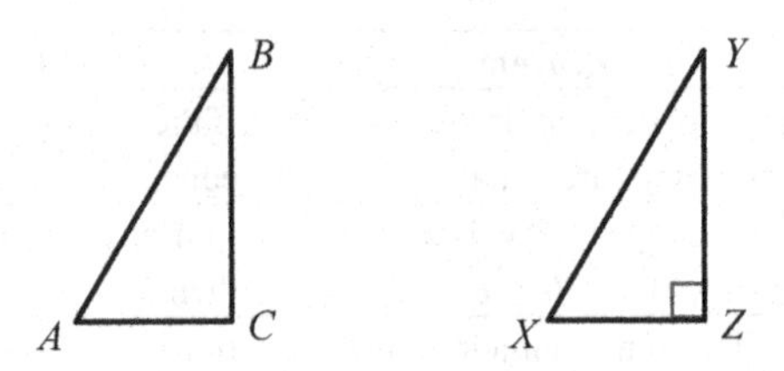

Figure 4.7. Proof of Converse of Pythagoras' Theorem

4.4 *Zero Product Rule (Zero Factor)*

This rule is used to solve quadratic equations by factorisation, but its justification is rarely covered in textbooks. A common attempt to justify this rule is to state that multiplying a number by 0 results in 0; this is actually the converse of this rule. The deductive proof is actually quite straightforward and could be easily taught:

- If $ab = 0$ and $a \neq 0$, then divide both sides by a: we get $b = 0$.
- Repeat the above if $b \neq 0$. This leads to $a = 0$.

An intuitive feel about the zero product rule can be gained by a simple thought experiment. Imagine a rectangle of sides a and b. Its area is ab. Mentally shrink this rectangle so that its area becomes zero. This happens only when one side or both sides become zero.

Students often make this wrong generalisation: If $ab = c$ and $c \neq 0$, then $a = c$ or $b = c$. This can be readily shown to be false by numerical substitution.

4.5 *Axiomatic system*

Mathematical deduction covers the following features.

- *Undefined objects*, such as points, line, sets, numbers.
- *Axioms*; these are propositions taken to be self-evident and true without proof; for example, one of Euclidean's axioms states that "one can draw a straight line segment by joining any two points." When the set of natural numbers is developed under Peano's postulates, one axiom is that "every natural number has a unique successor."
- *Conjectures*; these are propositions believed to be true but the proofs have not been found to date; they remain the *unsolved* problems for the time being. A well-known example is the Goldbach's conjecture: every even integer greater than 2 is a sum of two primes, e.g., $100 = 3 + 97 = 29 + 71$. This is often set as an investigation to introduce students to the idea of conjectures, but there is no expectation that they will be able to prove this! Many mathematics educators encourage students to formulate their own conjectures and try to prove or disprove them. This gives them the opportunity to think like mathematicians, which should be a major goal of mathematics education.
- *Theorems, formulae*; these are propositions that have been proved to be true. Some famous theorems are named after the main mathematicians who discovered or proved them, but most are not so labelled. *Lemmas* are valid propositions used as stepping stones to prove a more important theorem, and *corollaries* are

propositions that follow almost directly from one that has just being proved.

Other important ideas about deduction to be taught in mathematics lessons are noted below.

- *False* statements may be disproved by finding only one counter-example or using other methods of proof.
- *Impossible* problems are those that have been proved to have no mathematical solutions. They are different from conjectures in that the latter has not been proved or disproved. Famous impossible problems include trisecting a (general) angle or to square a circle using only compasses and straightedge and solving the general quintic equation in a finite number of the four operations and root extractions. The proofs that these problems cannot be solved under the given conditions are beyond most school students, but discussing these impossible problems can fire their curiosity about what mathematics can or cannot do.
- *Puzzles* or *riddles* are problems that do not have *obvious* solutions. These can be used to challenge students to think creatively or to enjoy thinking for its own sake. Examples include Towers of Hanoi, magic squares, card tricks with mathematical explanations, and Sudoku.
- *Fallacies* are arguments that appear to be true but are not, and finding the flaws in fallacies can foster reasoning and deepen mathematical understanding. Well-known examples are purported *proofs* to show that 2 = 1, -1 = 1, every triangle is isosceles, and so forth; see the comprehensive collection of fallacies and riddles compiled by Northrop (1944).
- *Paradoxes* are statements that appear to contain contradictions that might be true or are logically impossible, such as the liar paradox (*This statement is false*). A well-known paradox in statistics is the Simpson's Paradox: the same trend in different groups is changed when the groups are combined (Lesser, 2001). A realistic example with fictitious data of this paradox is shown in Table 4.9. Discovery method produces higher percentage pass than Direct

Instruction in each of the two studies, but the conclusion is reversed when the data from both studies are combined.

Table 4.9
Simpson's paradox

Study	Treatment	Sample	No. Passed	Percentage Pass
A	Discovery	400	120	30%
	Direct instruction	80	8	10%
B	Discovery	60	30	50%
	Direct instruction	500	200	40%
Combined	Discovery	460	150	32.61%
	Direct instruction	580	208	35.86%

The full axiomatic system is rarely taught nowadays in schools, especially in countries that have embraced the *Down with Euclidean Geometry* reform during the *Modern Mathematics* era in the 1970s. In these systems, Euclidean geometry has been replaced with transformation geometry or totally removed. Nevertheless, geometry proofs without axiomatic foundations are still taught, sometimes in a sequence of several related theorems. The purpose is to inculcate a sense that mathematics theorems are interconnected in a hierarchy and a feeling that lessons about these theorems have temporal connections (Leong, 2012). An example of this approach is about several angle properties of the circle. The sequence of these properties is captured in the concept map shown in Figure 4.8. It is based on Volume III of Euclid's *Elements*[20] and this is typically taught at upper secondary level.

Euclid proved III.31 using exterior angles of a triangle, but a simple exercise is to deduce this from III.20; this will add a new link to the concept map. Under the inductive to deductive approach, students should first explore these properties through drawing and measurement before they embark on the proofs.

[20] http://aleph0.clarku.edu/~djoyce/java/elements/

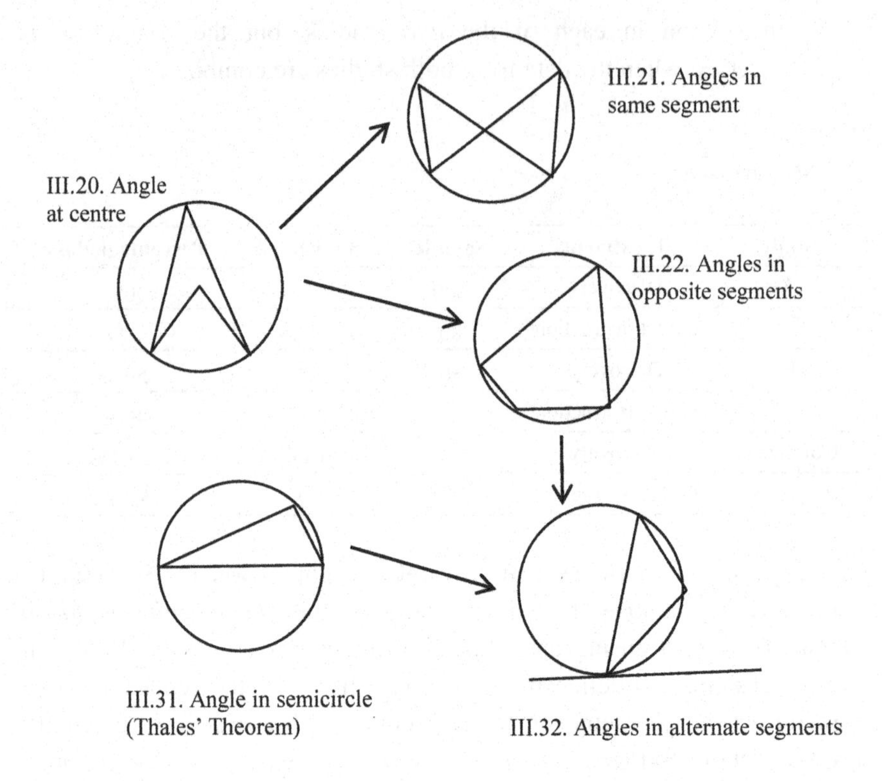

Figure 4.8. A concept map showing logical sequences about angle properties of the circle

The correct logical sequence of mathematical properties should be stressed to prevent students from producing proofs in the wrong order. The following is an attempt to prove the Pythagoras' Theorem.

Given triangle *ABC,* with *C* a right angle, $\sin A = a/c$, $\cos A = b/c$. Since $\sin^2 A + \cos^2 A = 1$, simple algebra shows that $a^2 + b^2 = c^2$.

Another wrong proof is to cite the Cosine Rule, $\cos C = \frac{a^2+b^2-c^2}{2ab}$, let $C = 90°$, then $\cos C = 0$, and simplify.

Both wrong proofs have been found in student work. Display the correct logical sequences in a concept map is an effective strategy to help students avoid making similar mistakes.

5 Acceptance of Results without Justifications

Up to secondary levels, some mathematical propositions have to be accepted on authority because both intuitive-experimental justifications and proofs are too difficult for students at these levels to understand. They should inform their students that mathematicians actually have the proofs. Some propositions commonly encountered that belong to this category are given below. As far as possible, teachers should know the proofs in case they are challenged by the mathematically gifted students.

- Surds such as $\sqrt{2}$ and $\sqrt{3}$ are irrational. The proof can be understood by the more advanced students.
- Rational numbers have terminating or recurring decimals. Use of calculator can be misleading; see Chapter 3, Section 2.
- Fundamental Theorem of Arithmetic: Every natural number greater than 1 has unique prime factorisation other than the order of the primes. This theorem provides the rationale why 1 is neither prime nor composite.
- There are an infinite number of primes. But whether or not there are an infinite number of pairs of twin primes (e.g., 2 and 3; 11 and 13) is still a conjecture.
- π is irrational. There is evidence that a sizeable number of teachers think that π is rational because they equate its value to ${}^{22}/_{7}$, without realising the rationale for the phrase *Take π to be* or the $\approx$ symbol (Wong, 1997a). This misconception is difficult to resolve because proofs that π is irrational are very complicated[21]. Since ${}^{22}/_{7}$ is rational, teachers and students can determine its recurring decimal. A more accurate approximation of π is ${}^{355}/_{113}$; its decimal repeats only after the 112th decimal place! This can be found using the *Excel* routine given in Chapter 6.
- e (Euler's number) is irrational. Again, the proofs are complicated[22]. To some students, the first nine decimal values displayed on a calculator (2.718281828) suggest a recurring pattern!

---

[21] http://en.wikipedia.org/wiki/Proof_that_%CF%80_is_irrational

[22] http://en.wikipedia.org/wiki/Proof_that_e_is_irrational

- e and π are called transcendental numbers, which are one type of irrational numbers, and the proofs for these two numbers appeared only in 1873 and 1882 respectively. However, it is still not known whether $e + \pi$, $e\pi$, or e^e is transcendental or not. These open questions show that mathematics propositions can be stated in simple terms but the proofs remain elusive. Exposing students to these mathematical realities can help them form more balanced beliefs about the nature of mathematics.

6 Heuristics

According to the Merriam-Webster online dictionary[23], *heuristic* is derived from the Greek verb, *heuriskein,* meaning *to discover.* In mathematics education, heuristics are considered general strategies that students may find helpful to discover the solutions when they do not have readily available methods to solve unfamiliar problems. They are rules of thumb, and using them does not necessarily lead to the correct solutions. The heuristic approach is made popular by Polya in his 1945 classic, *How to Solve It* (Polya, 1957), in which he explained several heuristics in dictionary style. Building on his model, the Singapore mathematics curriculum recommends teaching 12 heuristics under four categories; the 2006 version is given below:

- *Give a representation*: Draw a diagram, make a list, use equations.
- *Make a calculated guess*: Guess and check, look for patterns, make suppositions.
- *Go through the process*: Act it out, work backwards, before-after.
- *Change the problem*: Restate the problem, simplify the problem, solve part of the problem.

Given the strong emphasis on problem solving all over the world, the term *heuristic,* surprisingly, does not appear in some mathematics curricula (e.g., Australia, UK, and US), although suggestions such as to "consider analogous problems, and try special cases and simpler forms"

---

[23] http://www.merriam-webster.com/dictionary/heuristic

(CCSSI, 2010, p. 6) are obvious examples of heuristics. Naming these heuristics in curriculum documents alerts teachers to their importance and provides a common language for teachers to share their experiences of teaching problem solving strategies with colleagues. It is not known whether doing so actually improves the effectiveness of teaching problem solving, so it is worthwhile to investigate this.

Much has been published in Singapore and other countries on how to teach heuristics and to assess heuristic use (e.g., Chan, 2012; Fong, 1998; Foong, 2009a; Gojak, 2011; Kaur, 2008; Ng, 2002; Posamentier, Smith, & Stepelman, 2010; Toh, Quek, & Tay, 2008; Wong & Tiong, 2008; Yeap, 2008), and new studies continue to be reported in the literature.

Over the past two decades, many primary schools in Singapore have designed their own programmes to teach heuristics within the Polya's model. Some give memorable acronyms to their programmes, such as STAR (Study the problem; Think of a plan; Act on the plan; Reflect on the solution). This substantial effort is partly due to the fact that the PSLE examination includes *challenge problems*, to be solved using heuristics rather than learned skills. In secondary schools, however, the focus on heuristics is less evident because most of the problems in high-stakes public examinations require only content-specific skills rather than general strategies. Some schools, primary or secondary, also engage private companies or consultants to teach heuristics as enrichment activities outside formal curriculum time.

6.1 *Some local studies about teaching of heuristics*

It is not possible to summarise here the vast and ever expanding literature about problem solving built up over the past six decades worldwide. The following points gathered from the work of the MProSE team (Dindyal et al., 2012), Foong (2009b), Passmore (2007), and other researchers, point to several results worthy of further deliberation:

- Teaching of Polya-style problem solving is not yet common in mainstream mathematics lessons, especially at upper levels.
- There is some success in teaching problem solving at primary level, but mixed results at higher levels.

- It can become routinised if teachers adopt an expository style when teaching heuristics and problem solving.
- Students may become more motivated to master problem solving if the different stages of Polya's model are assessed and examinations require the use of heuristics.
- Competency in problem solving "develops slowly over a very long period of time" (Passmore, 2007, p. 51).

Many studies about problem solving deal with the whole Polya's model, and it is sometimes difficult to separate the effects of the whole model from those of individual heuristics. The purpose of the rest of this section is quite modest: to consider insights about heuristic use in the classrooms gathered from four local studies, arranged chronologically.

In the first study, Rajaram (1997) conducted a test of heuristics used by a sample of 73 Grade 8 boys enrolled in the Gifted Education Programme. These students were classified into *highly gifted* ($n = 19$), *averagely gifted* ($n = 30$), and *less than averagely gifted* ($n = 24$). They had learned various heuristics at lower grade levels, and a week prior to the test, they attended two 40-minute lessons to revise these heuristics. The test consisted of four non-routine problems to be completed in 80 minutes. The *Analytic Scale for Problem Solving* by Charles, Lester, and O'Daffer (1987) was modified to score the solutions. Six types of heuristics were identified: diagrams, list, pattern, algebra, re-state, and trial-and-error. The success of these heuristics varied with the problems and the level of giftedness of the students. A particularly salient finding was that the highly gifted students were able to recognise the content-specific mathematical principles of the problems and hence to apply this domain-specific knowledge to solve the problems, making few computational errors. In contrast, the other two groups were less able to do so, and they relied on general strategies and performed less well than the first group. The averagely gifted tended to use diagrams and listing, while the less than averagely gifted relied on trial-and-error. The findings are relevant to the issue of content knowledge versus general strategies in problem solving.

In the second study, S.O. Wong (Wong, S.O., 2002; Wong, S.O., & Lim-Teo, 2002) conducted a quasi-experimental study to investigate the effects of multiple heuristics (e.g., look for pattern, make supposition, systematic listing) on problem solving among a class of 39 Grade 6 students. Twelve 1-hour lessons about these heuristics were conducted over four weeks. Question prompts were provided to help the students when they were stuck, showed progress, or had found the solution. The last session of each week consisted of practice on a miscellaneous set of non-routine problems to develop heuristic selection and implementation. Performance in problem solving was assessed using a pre-test and a parallel post-test, each consisting of five non-routine problems. The answers were scored on appropriate use of heuristics and answers, modified from the same *Analytic Scale for Problem Solving* used by Rajaram. There was significant improvement in the test scores with very strong effect size (about 1.6), and the students were found to be more willing and flexible in using multiple heuristics after the training.

The third study covered both primary and secondary school students, and involved a sample of Grade 5 (n = 221) and Grade 7 (n = 64) students (Teong et al., 2009). They took two problem sets of parallel items about three months apart. Each problem set required 40 minutes to complete. In between the two administrations, their teachers attempted to implement the Polya's model. The heuristics used in their solutions to the two tests were analysed (Wong & Tiong, 2008), and five heuristics were found: systematic listing, guess and check, equations, logical argument, and diagrams. The success of each heuristic varied with the problems and students; however, for the given problems, systematic listing and guess and check were quite successful for both groups of students. Some students did not use the same heuristics to solve parallel problems across the two settings, raising the issue of consistency in heuristic use (Wong, 2008b). The reasons for this phenomenon are not known and require further study.

The fourth study to note here is called *Mathematical Problem Solving for Everyone* (*MProSE*), launched in 2009 and still on-going. The aim is to integrate problem solving into regular mathematics lessons at Grades 7 to 9. Its guidebook (Toh et al., 2011) covers the theoretical foundation of the project, a set of problems, *practical* worksheets, and lesson plans (10 lessons) to help teachers implement the Polya's model. The term *practical* is used to signify the practice of *doing* mathematics, similar to practical sessions in science lessons. Five participating schools reported their experiences with the use of these materials in a book edited by Leong et al. (2014). They discussed readiness of teachers to teach these new problem solving lessons, support from school leaders, adaptation of the packages (e.g., create a booklet of students' eureka moments), additional time to prepare lessons, and use of the MProSE assessment rubric. Qualitative comments were made about how some students had changed from initial focus on getting the right answers to subsequent attention to the problem solving process. However, quantitative findings about student performance are not reported.

To summarise, coding student solutions in detail, as used in the above studies, is time consuming but it can provide insights into the complex interactions between heuristics, problems, and students. This kind of information is useful for teachers, although research-wise, no definitive conclusion has been reached. Analysis of video clips about problem solving can also provide useful insights at fine-grained levels (e.g., Schoenfeld, 2008; Teong et al., 2009). Further experimental studies like those by S.O. Wong and MProSE should be conducted, since well-designed intervention programmes can inform teachers, curriculum designers, and teacher educators about how to help students effectively use heuristics. An even more ambitious research agenda is to compare the effects of domain-specific skills versus generic strategies in mathematics problem solving among the same groups of students, because the findings have implications for theories of mind, pedagogy, and curriculum design.

6.2 *Model drawing*

A distinctive and well-known feature of the Singapore primary mathematics curriculum is the *model drawing* (*bar model, model method*) method used to solve arithmetic word problems (Kho, Yeo, & Lim, 2009). Note that the term *model* in this method has different meaning from *mathematics model,* which is "a mathematical representation or idealisation of a real-world situation" (Ministry of Education, 2012, p. 18).

Model drawing is considered by many Singapore teachers as *the* heuristic, which is related to *drawing a diagram*, but the official document places it under different arithmetic topics and notes that it "serves as a bridge between arithmetic and algebra" (ibid, p.33), especially at the early stage of algebra learning at Grade 7. In a recent interview (Teng, 2014), Kho explained that his team developed this method in the 1980s based on Bruner's three stages and Greeno's approach of part-whole relationships. In other countries, this method comes under different labels, such as *strip diagrams, link diagrams,* and *thinking blocks*[24] (e.g., Ferrucci, Yeap, & Carter, 2003; Forsten, 2010; Hoven & Garelick, 2007; Xin, 2012; Walker, 2011). It is noted as one of the key practices that can account for the high performance of Singapore Grade 4 pupils in the TIMSS studies.

This method is now taught from Grade 2 upwards, and almost all schools have their own programmes about model drawing. Private tuition centres also offer model drawing classes for students and parents. Many assessment books and articles have been written about this method (e.g., Fong, 1993; Yeap, 2010).

The method involves drawing *bars* (*boxes, units*) of different lengths to represent the quantities given in an arithmetic word problem. It may be divided into parts according to the given conditions, but the lengths of the bars do not need to be proportional to the quantities, only relatively correct. The bars are then labelled with the respective quantities, and comparisons are made. The idea seems straightforward, but its power arises from the following aspects:

- The pictorial representation of the problem and the configuration of the bars embody the underlying mathematical structure of the problem. This helps students to better see the relationships among the given quantities and thereby compute the required values using standard operations.
- The labels and relative sizes of the bars provide external working memory space to alleviate extraneous cognitive load of the task.
- The method is systematically repeated and extended under different arithmetic topics: natural numbers, fractions, percentages,

---

[24] http://www.ThinkingBlocks.com

and ratios. There are three basic types of models, which can be used for these topics: *Comparison* (*difference*) model, *Part-whole* (*part-part-whole*) model, and *Before-after* (*change*) model. Through this systematic approach spread over four to five years of learning, students can use this technique to solve a variety of difficult word problems.

- Alternative models can be drawn to solve the same problem. This encourages creative thinking.
- Online practice is now available through the *AlgeTools* software[25] designed by the Singapore Ministry of Education and the National Institute of Education.

The example below shows how alternative models of the same word problem may be used. It uses the comparison model taught at Grade 4.

Problem. *Meiling and Sheela spent 220 hours at the children's home. Sheela spent 100 hours more than Meiling at the children's home. How many hours did Meiling spend at the children's home?*

Model 1. The typical approach is to treat Meiling's hours as the basic unit. The steps are shown in Figure 4.9. Meiling spent 60 hours at the children's home. Students should check this answer against the original conditions.

Model 2. This second model is less obvious than model 1 above. It treats Sheela's hours as the basic unit. The steps are shown in Figure 4.10. At the second step, one unit of Meiling's hours is added to 100 hours to make up another unit of Sheela's hours. The rest is self-explanatory.

[25] http://algetools.moe.edu.sg/

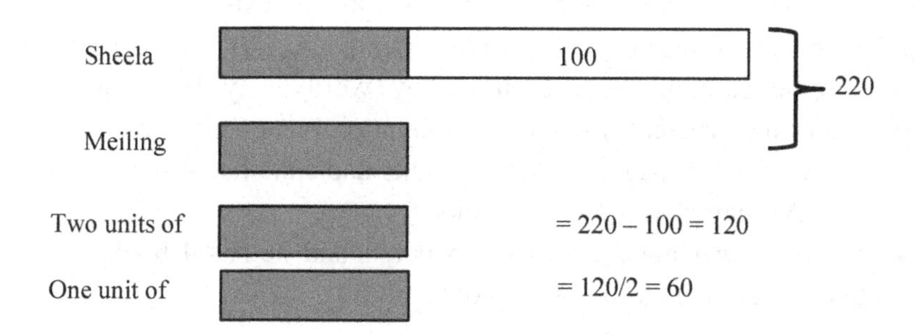

Figure 4.9. Model drawing of a problem: Meiling

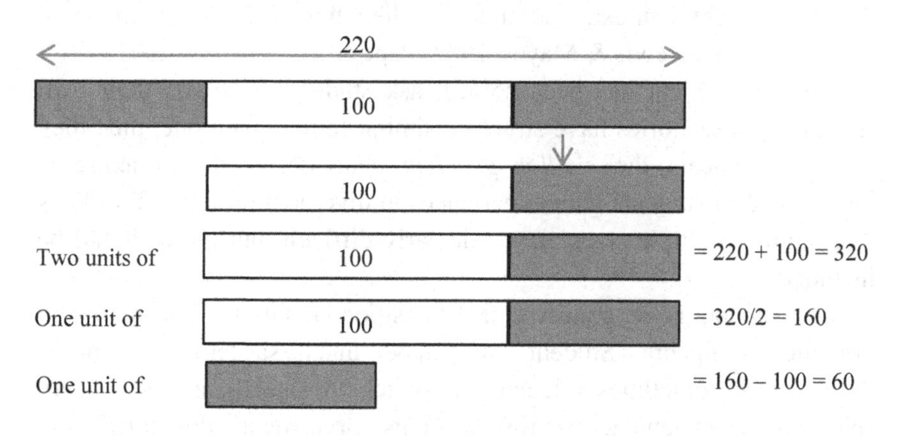

Figure 4.10. Model drawing of a problem: Sheela

Experiences of teaching model drawing over the past three decades have surfaced several teaching and learning issues. Two issues are discussed below.

The first issue is that students draw wrong or incomplete bar diagrams because their poor English literacy prevents them from understanding the problem context. They also do not know what the problem is asking them to find. One possible remedy is to ask them to repeat the problem in their own words and to listen carefully to what they are saying; for example, some students may say "100 hours" instead of "100 hours *more*." Another strategy is to engage them in interpreting the problem using concrete manipulatives starting with small numbers. Personalising the problems with students' names and daily experiences can motivate them to want to make sense of the problem contexts. However, the popular key-word approach (e.g., *more* means to *add*) should be used with extreme caution because these key words may be used in a consistent or inconsistent way under different problem contexts:

- *Consistent* language: Ali has 5 books and Jane has 4 books *more* than Ali. Jane has $5 + 4 = 9$ books.
- *Inconsistent* language: Ali has 5 books and he has 4 books *more* than Jane. Jane has $5 - 4 = 1$ book.

Research has shown that students consistently perform poorly under the inconsistent context, and this is called the *inconsistent language hypothesis* (e.g., Lewis & Mayer, 1987; Pape & Tchoshanov, 2001).

After a problem has been solved, ask students to create their own stories. If these stories have structure similar to the given one, then they help to inculcate the *similar problem* heuristic. If the structure is different, then students gain experiences in divergent thinking. Problems that require multiple steps are particularly difficult, but these should be included.

The second issue concerns the transition at Grade 7 from model drawing to algebra. Students who have much success with model drawing are sometimes reluctant to switch to algebra, and secondary school teachers tend to penalise students for using model drawing to solve algebraic word problems involving linear equations in one or two variables. The intended curriculum recommends using model drawing to bridge arithmetic and algebra.

One successful strategy is to let students first solve the problem using model drawing and then represent the *units* in the model drawing by variables. In Figure 4.9, let x be the number of hours spent by Meiling. The equation is $x + (x + 100) = 220$. In Figure 4.10, let y be the number of hours spent by Sheela. The equation is $2y = 220 + 100$. Once again, stress that alternative algebraic equations can be used, depending on which unknown is assigned. Let students discuss the efficiency of the algebraic method vis-a-vis model drawing. Sooner or later, the students will encounter problems that cannot be solved using the model approach, such as those with quadratic equations.

Research into model drawing has generated insights that can be translated into practice. Yeap (2010) provided information about how well students could use different models and which ones they preferred. Other studies examine students' drawings in detail in order to understand the underlying thinking. On the types of errors made in model drawings, Goh (2009), Poh (2007), Ng and Lee (2009), and Yan (2002) produced different classification schemes, which vary according to the topics, grade level, and cognitive demands of the problems. Lee and Ng (2009) used functional magnetic resonance imaging (fMRI) to show that algebraic method demands more working memory resources than model drawing, but it is not easy to match word problems to the loads on working memory. Primary school teachers are also keen to study model drawing through school-based action research. Under the aegis of the *North Zone School Based Curriculum (SBC) Communities of Practice (CoP)*, ten primary schools have conducted collaborative action research projects on model drawing, and their reports can be found in the North Zone series called *Celebrating Learning through Active Research (CLEAR)*. In these studies, scores were given for correctness of the models as well as the final answers, a common practice in school tests.

7 Question-and-Answer (Q&A)

A highly recommended strategy to help students sharpen their mathematical reasoning is to engage them in active question-and-answer (Q&A) during whole class or face-to-face interactions. Q&A is a well-established routine used in teaching, mathematics and other subjects.

The typical Q&A format is called *Initiate-Respond-Feedback* (*IRF* or *IRE, E* for *Evaluate*). The teacher *initiates* the Q&A routine by asking the class a question, calls on some students to *respond* to the question, and then *evaluates* the responses, and gives *feedback,* typically *yes, no,* vague praise, or brief comments. The vast literature on questioning reveals several shortcomings of IRF (e.g., Chapin, O'Connor, & Anderson, 2009; Henning, 2008; Rowe, 1978, 1987; Stahl, 1994; Walsh & Sattes, 2005; Wragg & Brown, 2001). They are:

- Most of the questions asked by teachers are low order in cognitive demand, such as recall of facts, next step in learned procedures, or answers to basic computations.
- Little time (less than one second) is given for students to think before they respond. This time for thinking is called *wait time* (*think time*). Similarly, teachers respond almost immediately to the answers given by the answerers, without providing time for the answerers to be ready to receive feedback and for other students to think about the given answer. Research shows that including these two types of wait times improves the quality of answers and discussion as well as classroom behaviours.
- Teachers seldom probe thinking behind wrong answers.
- Teachers tend to focus on target students who often volunteer to answer questions or are high achievers, ignoring the other students.
- Call on an answerer before asking the question; the reverse sequence is better to get more students to think about the question.
- Teachers tend to get response from only one student per question, instead of opening the question up for discussion.
- Very few students ask questions during lessons. This will be taken up in Chapter 8 on metacognition.

These shortcomings can be overcome by enhancing IRF with findings about classroom questioning, feedback, and formative assessment (e.g., Hattie, 2009, 2012; Wiggins, 2012; Wiliam, 2011; Wood, 1998). Table 4.10 shows how to enhance IRF by adding several options culled from the literature. These suggestions are not exhaustive. In the table, technical terms used to describe classroom questioning are in italics, and teachers should be familiar with them and use them in professional dialogues about classroom events.

Table 4.10

Enhanced IRF (Initiate-Respond-Feedback)

Basic IRF	Considerations	Enhanced IRF
Teacher asks question.	Students understand the question; surprisingly many students do not!	• Begin Q&A routine with, "Now I am going to ask …" • *Thumb up*: Ask students to show thumb up if they understand, thumb down if do not. • *Repeat*: Call on another student to repeat question.
	Time to think of an answer.	• *Wait time 1*; 3 s to 10 s. • If the question requires longer thinking time, change to *discussion* mode. • Maintain eye contact with the class.
Call on one student per question, usually target students.	Not all students are engaged.	• Call on a few students at random to answer the same question. • Keep a record of those who have answered questions, so that over several lessons, all students in the class have done so.
Student responds.	Students tend to mumble answers, not audible to others.	• Train students to speak clearly in class. • *Repeat;* ask student to repeat answer or teacher to repeat for student (not recommended) • *Re-direct* to another student: Can you repeat what … has said? • *Re-voice*: teacher repeats student answer to seek clarification: So you are saying …
	Time to think about the given answer.	• *Wait time 2;* about 10 s.
	Gather responses from other students.	• Thumb up routine to find out what other students think about the given answer. • Put up student's answer on the board.
	Student is stuck.	• For the answerer; funnelling, use low-level questions, shorter steps, or hints leading to expected answers. • Re-direct question to other students. • Think-Pair-Share; students write down their answers. • Come back to the student after further explanation to the class.
Teacher evaluates and gives feedback.	Many options	• See Table 4.11.
Move on.	Do not get stuck at one question for too long.	• Give and explain the answer. • Maintain momentum of the lesson.

The purpose of giving feedback is to narrow the gap between the student's current understanding and the expected standards. Effective feedback, oral or written, should be

- prompt and timely, to closely follow the student's answer or work;
- specific, so that the student can do something to improve their learning;
- encouraging, to avoid commenting on too many shortcomings; and
- easily understood by the target student.

During Q&A, teachers must be able to judge whether the answers are right or wrong and to give meaningful feedback on the spot. This requires strong mathematics knowledge and thoughtful planning to anticipate potential wrong answers. Different types of feedback can be given to right or wrong answer. Some suggestions are given in Table 4.11, illustrated with answers to the simple question: What is 56 × 3?

Table 4.11

Feedback to student answers to the question, "What is 56 × 3?"

Answers	Principles	Samples of Feedback
Correct answer: 168	Acknowledge Praise effort.	• Good, ok. • Your effort is appreciated.
	Seek reasoning.	• Explain how you get the answer. • Does the answer make sense?
	Find out what other students think.	• Ask a few more students. • Which method is efficient?
	Reinforce the rule; extend it.	• Ask several similar questions and direct them to students who have not been called on.
Wrong answer: 1518	Give answerer another attempt.	• It is too big. Think again. Wait for the same answerer to respond.
	Seek reasoning.	• Explain how you get the answer.
	Re-direct question to other students.	• John, what do you think? This could lead to discussion.
	Funnelling, remind student of previous work.	• Remember how we line up the numbers?
	Divide into smaller steps.	• What is 6 times 3? What do you do with the "carry over"?
	Do not get stuck for too long.	• Explain the solution and move on to other examples.

8 Mathematics Discussions

Discussion (*discourse, student talk*) is an extended form of Q&A between teacher and students or between student and student. It provides invaluable opportunities for students to clarify their ideas and solutions, to learn to communicate their mathematics ideas, and to become metacognitive because they can compare their own thinking to that of others. For the teachers, discussion is an engaging, also somewhat stressful, way to immediately monitor students' understanding and to the spot.

Learning to conduct fruitful discussions during lessons requires much deliberate practice. First, learn to design tasks that allow for different solutions to reach the same result or that have different answers. For example, after teaching how to derive the area of a triangle in the canonical form (see Figure 4.11(a), where the base is horizontal and the opposite vertex is *within* the base), let students discuss what happens when it is in non-canonical forms as in Figure 4.11(b) and (c). Students may work on this task either individually or in groups before whole class discussion commences. There is misunderstanding among some teachers that that discussion must be based on group work.

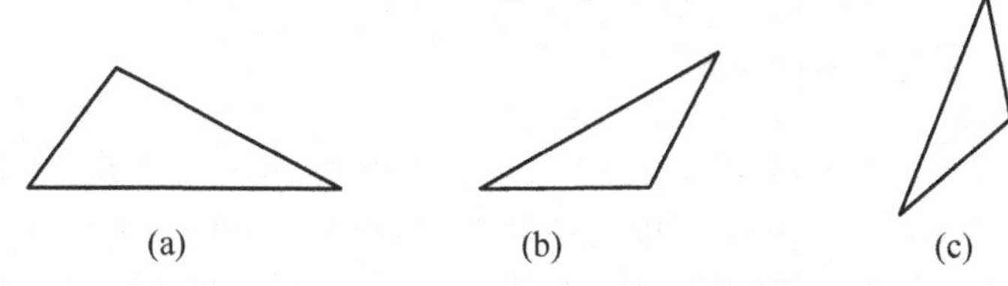

Figure 4.11. (a) Canonical form (b) Horizontal base (c) Slanting base

Second, master the more advanced Q&A techniques, in particular, funnelling, re-voicing, re-directing, and finding out how many students agree or disagree with a proposed solution using thumb up and voting.

Third, be aware of the different structures of organising discussion. Hintz and Kazemi (2014) described two structures to serve different purposes of mathematical discussion:

- *targeted sharing* to lead students to arrive at a consensus about particular concept or rule; and

- *open strategy sharing*, when students explore different ideas or solutions to solve the same problem.

Krpan (2013) described other structures: think-pair-share-write, talk prompts, clipboard walk and talk, mathematical clothesline, and *Bansho* (a Japanese strategy to let students place their solutions on chart papers in front of the class).

Students need to appreciate that discussion is a powerful learning experience that complements paying attention to teacher explanation and demonstration. Studies by Lubienski in the US (cited in Gates & Noyes, 2014) suggest that students from different socio-economic status (SES) background interpret discussion activities and open-ended problem solving differently. High SES students "thought discussion activities were for them to analyse different ideas whilst low SES students thought it was about getting right answers" (p. 45). High SES students were found to try harder to solve open-ended problems, while low SES students became easily frustrated with these problems. Teachers need to find out whether their students display similar tendencies, and if so, to convince them of the benefits of engaging in discussion.

9 Reasoning Questions

Students can sharpen their reasoning and heuristic use if they frequently ask themselves the following questions, after they have answered similar questions posed by their teachers using the (you) versions:

- How do I (you) get this?
- Why does … work here?
- How do I (you) convince myself (yourself), my (your) friends that this is true?
- Why do I (you) think … is *not* correct?
- Can I (you) use diagram/word/symbol/… to show my (your) thinking?
- Why did the teacher ask that question?

The last question in the above list has puzzled some teachers who attended my course on Q&A. It is really a metacognitive question for the students. It discloses to them that there is a purpose behind teacher questions, whether they are about concepts, rules, or processes. Being aware of this purpose leads them to better appreciate the tasks they are asked to work on and to minimise the unhelpful responses identified by Lubienski as cited above.

10 Concluding Remarks

According to Sawyer, "[t]he essential quality for a mathematician is the habit of thinking things out for oneself"[26], and this applies to mathematics students too. Bruner (1964b) emphasised the same point in this way:

> We teach a subject not to produce little living libraries on the subject, but, rather, to get a student to think mathematically for himself [sic] … Knowing is a process not a product. (p. 335)

Sawyer also pointed out, as quoted in the epigraph of this chapter, that one needs to revise one's reasoning when the wrong conclusions are obtained. This habit of mind is best learned through applying mathematical reasoning and heuristics to solve different types of problems and reflecting on the experience of doing so. This chapter has outlined how students can be aided to progress from intuitive-experimental justification to formal deductive proof in their mathematical journey. The enhanced IRF, discussion, mapping out logical sequences for a series of lessons about a set of mathematics properties, and posing questions about one's reasoning are some of the effective strategies to promote this progression. Eventually, the students must be able to apply their reasoning to solve and model everyday problems. This will be taken up in the next chapter.

[26] http://www.wwsawyer.org/sawyer-quotes.html

Chapter 5

Applications: View the World Through Mathematical Lenses

The ability to use mathematics in daily life is probably the most widely cited goal of mathematics education all over the world. This prepares students to apply mathematical knowledge to make decisions about daily situations, to reason about claims made based on quantitative or graphical data, and to support future learning of mathematics and other subjects. In short, students develop the aptitude to view the world through mathematical lenses and at the same time to gain new world knowledge about their environment. Two models are explicated in this chapter: one adds context knowledge to the Singapore mathematics curriculum framework and the other embeds *National Education* into the mathematics curriculum. Mathematical modelling is discussed as an extension of real-life applications. The given examples illustrate how these two models can be used to achieve the application goal in mathematics lessons.

Knowing is not enough; we must apply. Johann Wolfgang von Goethe (1749 – 1832)

1 Query about Mathematical Applications

Many students are eager to know why they have to study mathematics. This question about purposes or goals can be answered by citing the ideas given in Table 1.3 of Chapter 1. At the top of that list is daily application of mathematics. Students need authentic examples to answer their persistent query about "When am I going to use this in my life?"

Application of mathematics to daily situations is probably the most widely cited goal of mathematics education all over the world. They are also called by many different labels: *real-life* applications, *real-life problems, problems with real-life contexts, realistic problems, authentic tasks*, and *everyday mathematics*. There are subtle differences in what they mean to different educators; for example, *realistic* is artificial and not really *real.* But they all share the essential need to distinguish these problems from academic mathematics (Brenner & Moschkovich, 2002). To clarify the idea of authenticity of PISA items, OECD (2009) provided a more focussed perspective by differentiating between application and practice:

> this use of the term authentic is not intended to indicate that mathematics items are genuine or real. PISA mathematics uses the term authentic to indicate that the use of mathematics is genuinely directed to solving the problem at hand, rather than the problem being merely a vehicle for the purpose of practising some mathematics. (p. 92)

For brevity, all the above terms will be used interchangeably in this book.

There are four areas in which mathematics can be applied:

a) Apply specific skills to one's own daily situations; see Section 3.
b) Apply mathematical processes to diverse contexts; see Section 4.
c) Apply mathematics in the study of other subjects; see Section 5.
d) Apply mathematical attributes, such as curiosity, perseverance, and attention to details, to daily situations.

The next section introduces a fifth area of application called *context knowledge,* which has not been as adequately covered in the intended curriculum as it should be.

2 Context Knowledge

A critical goal of school education is to arouse students' interest in the physical and social world around them and to gain new knowledge about that world in order to have a meaningful life. Every school subject should contribute to the attainment of this important goal.

Mathematics can play an important role here. Mathematics problems should provide opportunities for students to learn something new about the problem contexts. This will enrich their life experiences, enabling them to understand their environments through mathematical lenses and to make informed decisions whenever they are confronted with similar problems in their *real,* day-to-day situations. Conversely, students need real-life knowledge in order to interpret the contexts or *story lines* of word problems. This kind of knowledge that comes from outside mathematics is called *context knowledge, real-world knowledge* (Ernest, 1998), or *factual knowledge* (Mayer, 2006a).

To emphasis this type of knowledge, I had included it as an extension to the Singapore Pentagon framework (Wong, 2008a); see Figure 5.1. In this extension, context knowledge serves two purposes: it is a factor of successful problem solving and it is a desirable outcome of solving problems with real-life contexts.

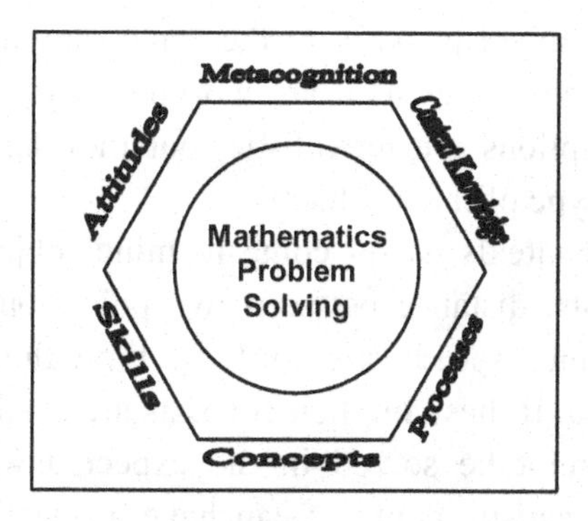

Figure 5.1. An extended mathematics curriculum framework with Context Knowledge

Examples of context knowledge as a learning outcome will be given in Section 3 onwards. The rest of this section examines three ways in which context knowledge appears in problem solving: *neutral, implicit,* or *conflicting*.

Neutral contexts. The structure of the problem remains unchanged when the context is changed. Example:

> Barry has three pencils and Wendy has four pencils more than Barry. How many pencils do they have altogether?

Changing the names of the persons or the object does not change the problem structure, so both contexts are neutral. Students do not need to apply substantial context knowledge to solve such problems, and neither do they learn new things about the world. Many word problems are of this kind.

Implicit contexts. Many real-life problems have implicit assumptions about contexts, and students need to take note of them in order to be able to solve the problems. Consider this rate problem:

> 8 men take 6 days to paint a house; how many men are required to paint the same house in 4 days?

The men are assumed to work at the same rate and this rate is not affected by how many other men are working at the same time in the house. Both assumptions are unrealistic, but they have to be made in order to solve this type of rate problems.

Other implicit contexts easily come to mind: objects are treated as dimensionless points; distance between two points on a map is always along a straight line; speeds are uniform throughout a journey; air resistance is ignored. If these implicit contexts are challenged, then many word problems cannot be solved in the expected way. According to Shannon (2007), "realistic contexts can have a negative impact on the performance of students on mathematics tasks" (p. 178), but this should not prevent teachers from assigning such problems. They can enlighten students about the roles of these contexts by discussing questions such as: "How do these assumptions affect the solutions?"

Conflicting contexts. Although real-life contexts are given in word problems, they are often ignored so that the expected *mathematical* answers can be computed. A well-known example is:

> There were five ducks in a pond. A hunter fired a shot and killed one of the ducks. How many ducks were left?

The expected answer is 4, but real-life experience suggests none! In fact, an answer of zero is often marked wrong. After much exposure to this kind of problem where *real-world answers* are not accepted, students learn to ignore the contexts and think that mathematics answers do not need to make sense. This is called the *suspension of meaning* (Verschaffel, Greer, & De Corte, 2000). For instance, students may give answers such as 15.2 buses when the number of buses is asked for. Sometimes, real-life considerations may lead to the wrong answers. This is especially true for mean values. For instance, students may think that it is meaningless to give 4.6 (say) as the mean number of people in a family and they round it up to 5. In this case, they lack conceptual understanding of mean as a statistic.

Students need to learn how to handle these three roles of contexts in typical mathematics problems. More importantly, they must acquire new real-world knowledge after solving these problems. The following sections provide substantial examples of this approach.

3 Direct Applications of Specific Skills

The typical perception is that specific mathematics skills must be applicable to common, daily situations. These include counting the number of items in a set, calculating payments and changes, measuring heights and weights, and determining one's position in a cinema such as A5 for row A and seat 5 (cf. Bishop, 1991). These direct applications require only basic arithmetic operations with whole numbers, decimals, and percentages, usually learned in primary schools. In contrast, most of the secondary school topics, such as algebraic manipulations and Euclidean geometry, are hardly used in daily situations, if at all. Hence, it is difficult to answer the "when to use" question at upper levels with daily applications. Teachers at these levels must be prepared to admit this limited use and divert student attention to other types of applications.

Real-life mathematics tasks should use authentic data and scenarios. In addition to these two criteria, Zhao, Cheang, Teo, and Lee (2011) proposed four other criteria:

- Connect tasks to the curriculum.
- Assess multiple mathematics competencies and content knowledge.
- Scale tasks by different levels of difficulties.
- Provide enriching experiences to students; this is akin to gaining new context knowledge.

They applied these criteria to design extended mathematics tasks with real-life contexts for the *Singapore Mathematics Assessment and Pedagogy Project* (SMAPP) (Wong, Zhao, et al., 2012). The tasks cover 11 scenarios using Singapore data, including paper recycling, water usage, off-peak cars, and mobile plans.

The example below on sound loudness satisfies some of these criteria.

Title: Decibel problem

Context knowledge. Sound is measured in decibel and hearing loud music for many hours will result in hearing loss. Students can use this knowledge to prevent hearing loss by controlling the time they spend listening to loud music with headphones.

Problem. This item is suitable for Grades 7 to 8.

The loudness of sound is measured in decibels (dB). Noise from heavy traffic is about 85 dB and this can cause hearing damage if one is exposed to it for 8 hours or more. For every 3 dB over 85 dB, the exposure time before hearing damage occurs is decreased by half[27]. This information is used in the problems given in Table 5.1.

[27]http://www.noisehelp.com/noise-dose.html/

Table 5.1

Problems about hearing loss

Problems	Comments
a) If the noise is 88 dB, what is the exposure time before damage occurs?	Interpret statement for simple case.
b) John likes to listen to his music using ear-plugs at a high volume of 100 dB. How long could he do this before hearing damage occurs?	Use pattern or systematic listing.
c) Let T hours be the permissible exposure time for loudness L dB. Express T in terms of L for $L \geq 85$.	Translate verbal statement into algebra; idea of domain.
d) Draw a graph of T against L. Interpret the graph.	Relate equation to graph.
e) Find out more about the decibel levels of different daily events and the effects of noise pollution on hearing and other health issues, such as stress and anxiety. Spot noise-mitigating barriers along roads and railway tracks.	View sound-related events with a mathematical lens.

Textbooks sometime include the formula, $I = I_0(10^{L/10})$, where I is the intensity of a sound of L decibels and I_0 is a constant. Students are required to manipulate this formula, say, make L the subject of the formula or convert it into logarithms. This provides practice for skill mastery, but it does not help students gain new context knowledge about sound. The following extension aims to deal with this gap. This is suitable for Grades 10 to 12.

Context knowledge: Sound loudness is based on but different from sound intensity. Loudness is measured in terms of decibels (dB). An increase of 10 dB corresponds to a 10-fold increase in sound intensity. The reference level (basic standard) for sound intensity is $I_0 = 10^{-12}$ Wm^{-2}. In Table 5.2, the first item shows how to translate the above definition about loudness into algebraic form using pattern. This competency can be applied to other similar situations.

Table 5.2
Definition of decibel

Problems	Comments
a) Complete the following table.	Express pattern in algebraic form.

Loudness, L (dB)	0	10	20	30
Intensity, I (Wm^{-2})	I_0	$10\,I_0$		

Problems	Comments
b) Express I in terms of L and I_0 in exponential form.	Algebraic manipulation.
c) Does zero dB mean silence? Is it possible to have sound with negative decibel?	Interpret extreme values.
d) Develop a rule of thumb to relate sound intensities to sounds loudness.	Rules of thumb are used in daily reckoning. Linear rule is wrong here.

To further reinforce the competencies mentioned under the Comments column, other similar contexts can be used. For example, the Richter scale is defined in this way: An increase of 1 in magnitude in the Richter scale represents a 32-fold increase in the energy released by rock movements during an earthquake. By posing problems similar to those in Table 5.2 and solving them, students can understand reports about the size of earthquakes, which might affect their daily life. This approach highlights the relevance of mathematics when one reads news about the world (Paulos, 1995).

4 Applications of Processes

This type of application differs from direct application of specific skills by focussing on processes such as inductive and deductive reasoning, communication skills, and problem solving. According to Bertrand Russell, "What is best in mathematics deserves not merely to be learnt as a task, but to be assimilated as part of daily thought"[28]. These processes

[28] http://readbookonline.net/readOnLine/22896

aim to change how students think about the world and to raise their intellectual power. Two areas are discussed below: an *appreciation* of the mathematics underlying various properties and events and the *detection* of mathematical errors in daily situations. Some examples of these applications are given below.

Appreciation of underlying mathematics

- Students come to understand the reasons why median is preferred to mean in reports about salaries. For example, the median salary, in contrast to the mean salary, is less likely to be affected by extremely low or high salaries.
- To enlarge an A4 document into A3 size on a photocopier, one selects the enlargement factor of 141% instead of 200%, even though the area of an A3 sheet is twice that of an A4 sheet. This arises because for paper in the A-series its length is $\sqrt{2}$ times the width. As an application activity, ask students to check that this is true for A4 paper with dimensions, 297 mm by 210 mm, and then prove it using similar rectangles (Wong, 1989). This is a rare example of where a quadratic equation and similar rectangles are used to design everyday objects. Students can investigate whether the same property applies to other series of papers.
- Knowledge of probability should guard students against the *gambler's fallacy* (*fallacy of the maturity of chances*): when an event occurs less frequently than expected, it will occur more frequently in the future to achieve the expected balance. For example, after getting several heads in a row by flipping a fair coin, one may erroneously believe that more tails will come next in order to strike a balance among the numbers of heads and tails.

Error detection

- A newspaper article stated that the salt content in a local dish was 1,675 g. This error is due to two different ways of presenting the decimal point: comma in European countries or period in English-speaking countries.
- Compare the cross sectional areas of logs by their circumferences; for example, put a string around the log and use its length as a

measure of the area. When the logs have similar lengths, this method is used to compare the sizes of the logs.

- At a clearance sale with 60% discount, the shop offers an additional discount of say 20%. Total discount is 80%.
- A discount of 20% is offered storewide. A customer purchases two items. The cashier applies 10% discount to each item so that the total discount is 20%.
- Many bar charts found in print materials do not begin the vertical scale from zero, and this leads to misleading comparisons. A critical view about statistical graphs includes examining the sources of the data used to create the graph, the purpose of the graph, and any alternative interpretations. Wu (2005) created a 12-item 5-point Likert scale to investigate whether secondary school students held such critical views. For a sample of about 900 secondary students in Singapore, they were found to hold a moderate level of critical views about statistical graphs. They focussed most frequently on whether the graph is easy to understand (highest mean, 3.37) and least frequently on whether there is anything misleading in the graph (lowest mean, 2.86). In fact, "the students were more concerned about the presentation of a graph than its meaning and value" (p. 181).
- Wrong predictions are made due to faulty understanding of statistics, such as correlations, linear regression, proportion, and the Law of Large Numbers. These mistakes are found in the mass media and journals. A particularly alarming example from the US is that "By 2048, all American adults would become overweight or obese."[29]

Deeper conceptual understanding of mathematics will enable students to detect these errors and adopt a more critical view of how mathematical concepts and techniques are used to describe real-life situations. By focussing the mathematical lens onto these daily situations, they will better appreciate the applications of mathematics to their life.

[29] http://www.americanobesity.org/obesityForecast.htm

5 Applications across School Subjects

This is called the *cross-curriculum* (*cross-disciplinary, multi-disciplinary*) approach (e.g., Hansen & Vaukins, 2011; Maasβ & O'Donoghue, 2011) or the *integrated-environmentalist* curriculum strand discussed by Howson, Keitel, and Kilpatrick (1981). STEM (Science, Technology, Engineering, and Mathematics) and STEAM (adding Arts and Design) are recent initiatives of such integration to help students learn to apply scientific and mathematical reasoning and design principles to solve complex problems. Languages, music, physical education, and social studies have also been embedded into mathematics lessons, but to lesser extents. Students may also work in groups to apply their knowledge in several subjects to tackle a fairly extended problem under *problem-based learning* or the *project work* programme in Singapore[30]. Such integration is appealing in terms of providing holistic education beyond learning academic subjects in separate time slots, but many challenges remain. These challenges include:

- difficulty of getting different subject matter specialists in the school to work efficiently together on planning, teaching, and assessing substantial multi-disciplinary projects;
- lack of quality cross-curriculum projects that go beyond superficial integration;
- high cognitive demands of these projects on average and weak students;
- constraints of time and resources;
- difficulty in ensuring valid, reliable, and fair assessment of projects; and
- difficulty in differentiating assessment of mathematics knowledge from other competencies, especially language literacy.

Indeed, the authors of a recent review of the Australian mathematics curriculum recommended that "cross-curriculum priorities should not apply to the mathematics curriculum" (Donnelly & Wiltshire, 2014, p. 176). If such projects are implemented only once or twice per year, as is

[30] http://www.moe.gov.sg/education/programmes/project-work/

often the case in some local schools, there is insufficient opportunity for students to develop the competency to apply mathematics to other subjects and vice versa in a two-directional integration.

Despite these challenges, short cross-curriculum problems can be set as exercises to provide motivation and to enrich learning experiences. The rest of this section will include only science examples. According to Gauss, mathematics is the Queen of Sciences, and traditionally, there are strong links between these two subjects. Science lessons require mathematical skills such as graph plotting and algebraic manipulations, and these should be covered before the respective science lessons. What concerns us here is to use scientific contexts in mathematics problems.

Example 1. In computers, a bit is either 0 or 1. A byte consists of 8 bits. An integer is stored in a sequence of two bytes (or 16 bit). The first (leftmost) bit is the sign bit: 0 if the integer is zero or positive, and 1 if it is negative.

a) For example, the 16-bit computer representation of 5 is: 00000000 00000101. Write down the 16-bit computer representations of one hundred.
b) What is the largest positive integer that can be stored in this way?

Example 2. Scientific experiments often involve taking several measurements of the same quantity and calculating the mean. One way to estimate the uncertainty of the mean is as follows: for each value, calculate its absolute deviation from the mean; then take the mean of these absolute deviations. An object is weighed on four different scales, giving the values: 1.01 kg, 0.99 kg, 0.97 kg, and 1.02 kg. Apply this method to estimate its weight and indicate an error range.

Example 3. In October 1999, NASA lost its $125 million Mars Climate Orbiter because engineers from Lockheed used the Imperial units and those from the Jet Propulsion Laboratory used SI units. As a result, the orbiter came within 37 miles of the surface of Mars, whereas NASA required it to fly no closer than 87 miles. Express these distances in kilometres and estimate the percentage error involved (using the correct distance as the base).

Example 4. Environmental biologists often take aerial photographs of the region they are studying. To be distinguishable, a line of 100 m long in reality must be 2 cm long on the photograph. From what altitude must a picture of a region be taken if the camera has a focal length of 35 mm, given that $\frac{\text{photo dimension}}{\text{actual dimention}} = \frac{\text{focal length}}{\text{altitude}}$?

Example 5. Members of a family of chemical compounds contain Carbon atoms, C, and Hydrogen atoms, H. The first three members of the family are represented in Figure 5.2. Determine the formula for this family of compounds.

```
    H             H   H              H   H   H
    |             |   |              |   |   |
H — C — H     H — C — C — H      H — C — C — C — H
    |             |   |              |   |   |
    H             H   H              H   H   H
```

Figure 5.2. Pattern for a chemical compound

Example 6. The half-life of a radioactive material is 10 years. Write down its decay function in the form $N = N_o e^{-kt}$. How much of 200 g of the radioactive material will remain after (a) 20 years, (b) 1 year?

To be successful in this cross-curriculum approach, mathematics teachers must be keen to learn more about other subjects. They must also be curious about their own living environment: notice unusual events, read about them, pose questions, collect and analyse data, internalise the new learning, and so on. When they find learning new things fascinating and enjoyable, they will be able to better engage their students in similar learning through their enthusiasm and to serve as a role model of a life-long learner.

6 Mathematics-related National Education: A Singapore Initiative

In Singapore, the term *National Education* (NE) refers to the education policy that aims to develop national cohesion, cultivate the instinct for survival as a nation and instil in our students, confidence in our nation's future. It also emphasises on cultivating a sense of belonging and emotional rootedness to Singapore[31].

Over the past two decades since the launch of NE in 1996, many programmes have been implemented at the national and school levels. Events such as NE quizzes and celebration of the four core NE events, namely *Total Defence Day, International Friendship Day, Racial Harmony Day,* and *National Day,* are now regular parts of the school calendar. NE activities are to be embedded in lessons of different subjects, if appropriate; see chapters in Tan and Goh (2003). In 2007, the *Head, Heart and Hands* framework was introduced to engage students, educators, and the community about NE. For students, they use their *head* to understand the challenges facing Singapore; their *heart* to have a deep sense of belonging to Singapore; and their *hands* to contribute to Singapore's future. The NE messages are now included under the new Character and Citizenship Education (CCE)[32], first introduced in 2014.

For mathematics, the 2000 version of the curriculum recommended incorporating national and current issues as contexts in mathematics problems and encouraging students to discuss the implications of the answers obtained in relation to these issues. Popular issues include: history and geography of Singapore, expenditure and saving, consumption of water and electricity, and reduce-reuse-recycle. However, NE is no longer explicitly stated in the current version, but the approved textbooks contain NE-related problems.

Wong (2003b) proposed the *National Education* × *Mathematics Education* (NE × ME) framework shown in Table 5.3, as one way to stimulate discussion about how to infuse NE issues into mathematics lessons. This framework can be easily modified by educators outside of Singapore who wish to embed significant issues of their own countries into the mathematics curriculum.

---

[31] http://ne.moe.edu.sg/ne/slot/u223/ne/index.html

[32] http://www.moe.gov.sg/education/syllabuses/character-citizenship-education/

Table 5.3

National Education × Mathematics Education framework

		Mathematics Education (ME)		
		Contents	Processes	Culture
National Education (NE)	Environments: • Personal • National • Global	Standard skills using authentic data	Reasoning, communicating, investigating	History, philosophy, ethno-mathematics
	Values: • Personal • National • Universal	Implications	Implications	Perseverance, curiosity, logical thinking

The *environments* of a nation can be divided into three levels: the physical and social environments of individual students; the physical, economic, political, and cultural environments of the nation as a whole; and the global environment of which the nation is a member. Family and school environments may be subsumed under personal environment. The *values* in the framework also cover three types: those espoused by individual students; national identity or ideology; and universal values, such as compassion, kindness, and moral principles.

Mathematics education comprises three components:

- *Contents:* the problem requires the application of standard concepts and skills to real-life contexts and scenarios.
- *Processes:* the problem is non-routine and requires multiple skills, mathematical reasoning, communication, and investigation.
- *Culture:* the problem encompasses the philosophy, history, social impacts, literature, learning and teaching of mathematics, and ethno-mathematics (D'Ambrosio, 2006).

One may begin by designing an activity or problem from any cell and then moving on to other cells, if appropriate. Wong (2003b) described this process with 15 examples, which will not be repeated here. Instead, a new extended task is shown below. After solving the problems, students should discuss how the answers and contexts can impact their lives.

Title: Water use and crisis

Context knowledge. A sense of the amount of water expressed in cubic metres (Cu M, m^3); charge for water usage.

Personal environment

A person needs about 19 litres of water for daily consumption including drinking, showering, and cleaning. The water charge in Singapore is $1.17 per Cu M.

a) Convert Cu M to litres. Make a model of 1 Cu M to help students gain a sense of its capacity.
b) How much water is required per person per month?
c) Another estimate is that a person needs 4 to 5 gallons of water per day to survive. Compare this information with the above one.
d) A family used 22.4 Cu M of water in a particular month. What was the water bill for this family for that month?
e) Estimate the total amount of water consumed by your family per month. Give the answer in litres and cubic metres.
f) Calculate the water bill of your family based on the estimate in (e).
g) Explain any difference between the estimate and the actual bills over several months.

National environment

h) The national average of water usage for that month was 19.3 Cu M. Express the usage of the family in (d) as a percent of the national average.
i) Singapore launched the *10-litre challenge* in 2006 to encourage every resident to reduce daily water consumption by 10 litres[33]. Search the PUB website to find out how to meet this challenge. What methods can you and your family implement to help achieve goal?
j) The daily domestic water consumption per capita has decreased from 165 litres in 2003 to 153 litres in 2011[34]. Relate this information to the national average in (h) above.

---

[33] http://www.pub.gov.sg/conserve/Households/tenlitres/Pages/default.aspx

[34] http://app.mewr.gov.sg/web/Contents/Contents.aspx?ContId=1533

k) How much water will be wasted with a faulty (dripping) tap? Make a guess and carry out an experiment to verify your guess.

Global environment

l) According to United Nations Department of Economic and Social Affairs (UNDESA)[35], about 700 million people today suffer from water scarcity. Express this as a percentage of the world population.

m) Write a report about the measures taken to address this water crisis.

7 Mathematical Modelling (MM)

A mathematical model is a mathematical description or representation of the properties or functions of a real-world situation. For example, radioactive decay is described by an exponential equation, which constitutes the *model*. Mathematical modelling (MM) is the process of formulating the model. This process is typically described as a cyclic one: real world situation → mathematical model → mathematical solution → real world solution → real world situation (with new understanding). The thinking involved at the transition from one stage to the next includes: formulating, gathering data, solving, interpreting, and reflecting on the MM process itself. The goal to engage students in MM is twofold: to master the modelling process and to acquire new real-world knowledge, similar to gaining context knowledge in application problems.

Mathematical modelling was introduced in the 1970s as one type of extended problem solving. Over the past four decades, a vast literature has evolved internationally about different versions of MM, strategies to teach and assess MM, and findings about the impacts of these strategies on student performance in MM. Some published works are: Ang (2009); Berry and Houston (1995); Burghes et al. (1996); Cross and Moscardini (1985); Dindyal (2009); Houston et al. (1997); Huntley and James (1990); Kaur and Dindyal (2010); Lee and Ng (2015); Ng and Lee

[35] http://www.un.org/waterforlifedecade/scarcity.shtml

(2012); Stillman et al. (2013); Swetz and Hartzler (1991). The journal *Teaching Mathematics and Its Applications* publishes papers in MM at the upper secondary to tertiary levels. There are also the biennial *International Conferences on the Teaching of Mathematics Modelling and Applications* (ICTMA). Hence, there is no shortage of resources for teaching MM at school levels.

It is not possible to summarise this vast literature here, so the rest of this section will focus only on how to get started with a MM activity. There are two approaches to get started (Wong, 2002a):

a) Apply one's current knowledge of mathematics and the relevant domains to construct a mathematical solution; check the prediction of this solution against data; see examples under Section 7.1.
b) Collect data through experiments or surveys, create a model, and check its prediction, if appropriate; see examples under Section 7.2.

7.1 *Apply known knowledge*

In the secondary school curriculum, several mathematical techniques can be used to model specific real-life situations, briefly outlined below:

- Similarity of shapes: enlargement or reduction of shapes, photocopying, map scales.
- Quadratic graphs: projectile motions.
- Exponential functions: population growth, radioactive decay.
- Trigonometric functions: daily variations in temperatures, biorhythms, tidal waves.
- Probability: games of chance.
- Statistics: variations of a property in a population, e.g., IQ.

Knowledge of the relevant domains is also helpful. A plausible model is formulated and then validated against some data.

Example 1. Leaking bottle
This was a popular MM activity. Take a plastic bottle and make a small hole near the bottom. Fill it with water. What is the relationship between

the height (h) of the water above the hole and the time (t) that has elapsed since the hole was uncovered?

To derive a mathematical solution, one needs to know Torricelli's Theorem: Assuming negligible viscosity and streamline flow near the hole, the speed of the emerging water is given as $v = \sqrt{2gh}$ (Nelkon & Parker, 1982). Applying basic calculus, one can show that $\frac{dh}{dt} = -\left(\frac{A\sqrt{2g}}{B}\right)h^{\frac{1}{2}} = -kh^{\frac{1}{2}}$, for some positive constant k. Integrating, one obtains $h = \left(\sqrt{H} - kt\right)^2$, where H is the original water level above the hole. However, as the water level drops close to the hole, the quadratic fit is poor due to viscous forces and lack of streamline flow. When data were collected and plotted, as shown in Figure 5.3, the fit is close except near the end of the flow, as expected.

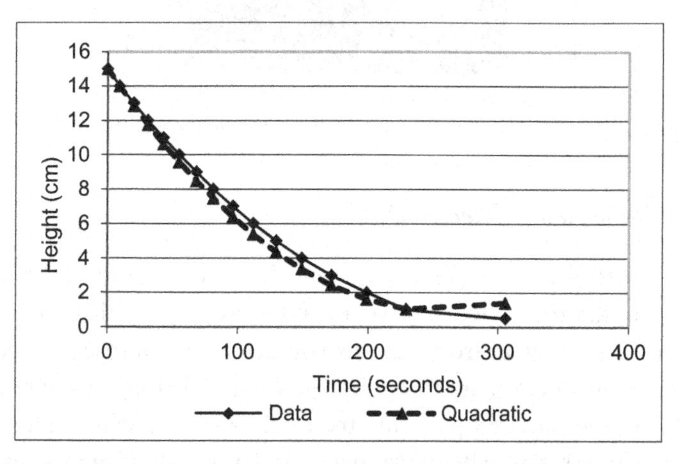

Figure 5.3. Graphs that model the leaking bottle problem

Example 2. Model the path of Merlion's spout
The Merlion is a well-known Singapore icon, and the MM activity is to model the path of the spout of the Merlion statue at the Merlion Park. The spout looks like a projectile, so a quadratic equation is called for. One solution is to take a photograph of the statue, paste it into Geometer's Sketchpad, and derive the equation using some suitable points. This is shown in Figure 5.4. The function, $6 - 0.06x - 0.04x^2$, provides a very good fit. Using ICT to tackle MM activities opens up new possibilities, and tech-savvy students can be motivated to do the mathematics.

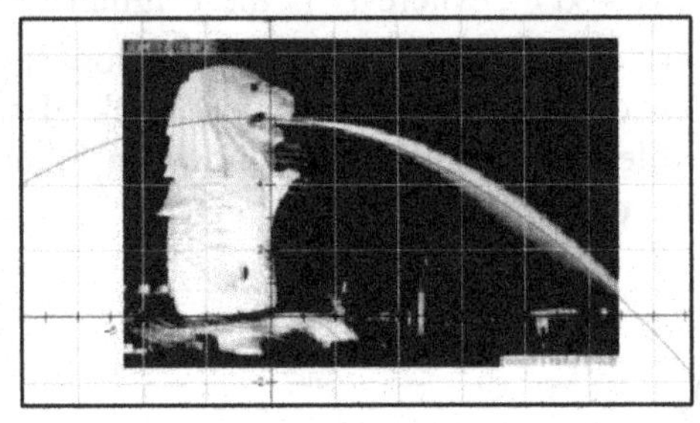

Figure 5.4. Modelling the Merlion's spout

7.2 *Models based on collected data*

Many MM activities assume that students lack knowledge of the relevant domains, so that the above approach is not feasible. In this case, students search for relevant data from various sources (to instil research skills) or collect their own data (e.g., through experiments or surveys), tabulate the data, plot graphs, notice patterns, try to fit the data with mathematical techniques (after the data have been transformed, if necessary), and finally check the validity of predictions based on the proposed model. This does not always result in good-fitting models, but it is the typical approach used by professionals when they attempt to analyse data without any prior theoretical knowledge. By working through MM activities in this way, students appreciate how the MM cycle is used to extend knowledge in a domain. This objective is more important than obtaining accurate models of the situations.

Example 1. Use of letters of the alphabet in text materials

There are no pre-determined rules about how frequent letters of the alphabet are used, which letters appear frequently as the first letter of a word, or what is the modal length of words in texts. To gain some insights into these questions, students can analyse randomly selected passages taken from various sources, and compare their findings with those reported by other researchers. For English, the top three letters used are: *E* (12.702%), *T* (9.056%), and *A* (8.167%)[36]. For Malay, a small student project found this: *A* (21.6%), *N* (10.4%), and *E* (7.6%) (Wong, 1979). These results can be used to decode Caesar cipher; a simple activity is as follows:

a) Create a message in English or other language.
b) Translate each letter a fixed number of steps, say *A* to *H*. This gives the coded message.
c) On receiving the coded message, determine the frequency distribution of each letter. Check this against the above findings. The most frequently used letter in the coded message is likely to be *E*. Decode the message using this translation and check whether it makes sense or not.
d) If the decoded message does not make sense, repeat step (c) with the next frequently used letter and so on.

Example 2. Clean up Singapore beaches and coastal areas

This MM activity raises students' appreciation that the same set of data may be modelled using different rules, and these rules may fit subsequent data differently. This demonstrates to students that real-life data are messy and do not always follow precise rules.

Singapore has been participating in the annual *International Coastal Cleanup* (ICC) event since 1992 to collect litter from beaches and coastal areas in Singapore. The number of pieces of items collected from 2002 to 2004 is given below[37]:

---

[36] http://en.wikipedia.org/wiki/Letter_frequency

[37] http://coastalcleanup.nus.edu.sg/aboutcleanup.html

Year	Number of pieces
2002	66 000
2003	74 000
2004	90 000

a) Model the amount of litter collected using a quadratic function. To simplify calculations, express the number of items in multiples of thousands.
b) Model the same information using an exponential function.
c) Graph the two functions on the same scale. Comment on the graphs.
d) Predict the amount of litter for year 2005 and 2014 using the quadratic and exponential functions.
e) The data for 2005 and 2014 were 108 000 and 174 714 respectively. Which function is a *better* predictor given the additional data? Explain the criteria used to define *better*.
f) Is there a limit to this litter growth?
g) Suggest ways to combat this problem at the personal, family, national, and global levels. Relate this to the *Values* component of the NE × ME framework in Table 5.3.

7.3 *From Additional Mathematics*

Additional Mathematics is a more advanced course for Grades 9 and 10. Its curriculum contains a major topic on transforming non-linear relationships into linear form to model a given situation. Typical questions follow this pattern:

a) Give a set of fictitious data about a real-life scenario, say object distance (x) and image distance (y) from a lens.
b) Ask students to transform the data according to a given rule, say $1/x$ and $1/y$, and plot the transformed data; this results in a straight line.
c) Deduce parameters from the line, and write down the modelling function.

Since students are given the expected transformation, they do not acquire the skill of looking at a plot of the data and exploring different transformations of their own choice. Since fictitious data are used, students are not expected to learn new context knowledge about the situation, and this misses a golden opportunity to help students appreciate that modelling is about *real* real-life situations. The following example is an attempt to avert these problems.

Example. Relationship between the orbital period of revolution (T) of a planet around the Sun and its mean distance (d) from the Sun
This activity shows how to derive a rule based on transformations of data. Ask students to search online sources for the relevant information, tabulate the data, plot them using *Excel*, and transform the data to arrive at a model for the relationship. Let them experiment with their own transformations and share their reasoning. This is Kepler's Third Law of Planetary Motion: T^2 is directly proportional to d^3 (published in 1619). The expected transformation is to take the square of T and the cube of d.

7.4 *Some teaching issues*

Mathematical modelling is a new component of the Singapore intended curriculum, and there is little research about how teachers cope with this activity. Ng (2010) described her experience of getting four groups of primary school teachers (4 per group) to work on the MM task to select two out of five women for a swimming event, given their swimming records in ten previous competitions. Three groups used average time as the only selection criterion, and one group produced an incomplete graph. She noted that the teachers were brief in communicating their reasoning, apprehensive about whether or not they were on the *right track*, and anxious that they could not directly apply an algorithm to tackle the task. In my work with Brunei and Singapore teachers, I have encountered the following three critical issues, which require some attention:

- Teachers with weak mathematics knowledge beyond what they are teaching and poor knowledge of domains outside mathematics only notice superficial features of the given data and graphs. This is evident from these two statements about the leaking bottle problem: "As time increases, the height decreases" and "The higher the hole is, the longer the time for the water to stop and vice versa." Thus, in order to *see* something deep in the data, one needs to *know* at least some knowledge of the phenomenon described by the data; see Chapter 4, Section 3.1.
- Some teachers confuse abstract mathematical models with physical objects. They believe that getting students to make models of furniture arrangement in a room or draw scaled diagrams of similar situations satisfies the requirement of including MM in their lessons. This is certainly not the intended purpose of MM in the curriculum.
- Among Singapore teachers, there is a confusion of MM with the model drawing method used to solve challenging problems. It is unfortunate that the word *model* is used to mean at least four different things: mathematical representation, model drawing, physical models of objects, and framework about mathematics education.
- Some teachers do not know how to design meaningful MM tasks. This is particularly serious among primary school teachers because their weak mathematics knowledge prevents them from simplifying MM tasks that use complex algebra and calculus for primary school students or designing their own tasks.

These observations suggest that teachers require training in both content (mathematics and other domains) and pedagogy. They need to enrich their own real-world experience through engaging in meaningful activities other than education. To be curious is certainly helpful.

In a similar vein, students require careful teaching to be able to complete modelling projects. Activities must be provided to help them bridge the gap between mathematics as learned in a neutral and symbolic form and mathematics to be used in a meaningful context. Finally, they are unlikely to develop robust MM skills if they complete only one or two MM projects per year, since MM is only a minor part of the curriculum. Innovative programmes are necessary. Parents can assist by providing novel life experiences even in mundane situations, such as cooking, sports, and shopping, so that their children are encouraged to see how mathematics is used to describe these experiences. The impacts of these suggestions should be investigated.

8 Application Questions

Students can learn to view their world through mathematical lenses when they constantly remind themselves to ask questions of what they read, hear, and notice about the use mathematical information in its numerical, graphical, verbal, and symbolic forms. Some pertinent questions to promote this critical view about mathematics applications and modelling are:

- Have you come across …in daily life?
- Give a real-life example of …
- Why do we study …?
- Is the link between … and … as reported correct? Can I reason this out?
- Do I understand how this works in the given real situation?
- What can I learn about the real world after solving these problems?

9 Concluding Remarks

Galileo Galilei (1564 – 1642) said: "The book of Nature is written in mathematical language." First, students must develop sharp observation skills to notice patterns and unusual features in data and scenarios around their environment. Second, they can apply appropriate mathematical techniques to solve problems related to these situations or to model new phenomenon. This chapter has provided many real-life situations that can be used in mathematics lessons. Both types of activity ought to increase students' context knowledge, as emphasised in the extended curriculum framework depicted in Figure 5.1, and to prepare them for further learning in mathematics and other subjects under cross-curriculum teaching. Third, Galileo also painstakingly designed new instruments to get more accurate results from his observations and to widen the scope of his observations. Likewise, ICT (Information and Communications Technology) now comprises many tools that teachers and students can exploit to deepen and broaden the contexts to which they can apply mathematics and to facilitate the teaching and learning of the subject. The following chapter will examine how this can be achieved when both teachers and students act as prudent masters of these ICT tools.

Chapter 6

ICT: Be Its Prudent Master

For the past six decades, Information and Communications Technology (ICT), involving extensive use of computers, the Internet, and digit devices, has revolutionised the practice of mathematics and its teaching and learning from pre-schools to universities. Numerous resources are now available to support different pedagogies, but research on the use of ICT tools in mathematics education has produced mixed findings. This chapter examines four major modes of ICT use and a computer-based assessment system. Given the pros and cons of using these modes for learning and teaching, teachers and students should strive to become prudent masters of ICT in order to achieve the intended curriculum goals.

Technology is just a tool. In terms of getting the kids working together and motivating them, the teacher is the most important.
Bill Gates

1 Introduction

Information and Communications Technology (ICT) or Information Technology (IT) is an umbrella term used to cover devices and applications associated with the computer. These include smartphones, tablets, Internet, intranets, app, video-conferencing, laptops, desktops, and new technologies that continually appear in the market. Older technologies such as radio, television, video-player, and calculator, now have digitised versions. In mathematics research, the emergence of ICT has changed some mathematics practices. The most significance change is the acceptance of computer-assisted proofs, after Appel and Haken proved the Four Colour Theorem in 1976, using more than one thousand hours of the computer time available at the time, although this can now be reduced to about an hour using faster computers (Stewart, 2009). Another advance is the ability to study complex systems such as fractals and chaos by using powerful graphing capabilities of new computer software and fast computer processing. These are examples where combining human ingenuity with ICT powers can produce complex results, somewhat like a *centaur*, which is a combination of chess player and chess-playing software. These changes in the mathematics discipline suggest that the contents and processes covered in the school mathematics curriculum should also change, but this has not happened in major ways in almost all the countries.

ICT use in education began slowly in 1970s and has since accelerated drastically due to a combination of these factors: enhanced hardware, lower cost, sophisticated software, teacher training and support, insights from research, and its growing use in assessment. Staunch advocates (e.g., authors of official curricula) and thoughtful critics (e.g., Healy, 1999; Oppenheimer, 2003) continue to debate the pros and cons of ICT use in education because research has produced mixed findings. These findings are based on old technologies and may no longer be valid for those in current use.

2 Aims of ICT Use in Mathematics Education

Several aims of using ICT in education, including mathematics, are given below.

- ICT tools can support different pedagogies, including direct instruction, guided discovery, self-regulated learning, and collaborative learning. Instead of adhering to a particular pedagogy, say inquiry learning, teachers can use ICT tools in different ways to offer enriched learning experiences to their students.
- Students can transfer ICT skills from the classrooms to daily situations. However, ICT tools rapidly change in features or

become obsolete, so ICT lessons should not be too rigidly constrained by current technologies and usage. Adaptive and mindful use of diverse ICT tools is more desirable than technical competence.

- ICT activities can promote new literacies, such as image literacy, data literacy, and search literacy, which will become more important in the future. These correspond in mathematics education to skills of modelling, visualising, simulating, searching for information, conjecturing, justifying, and working on large dataset.
- Some complex mathematics can now be introduced in schools because the necessary computations can be done using software so that the focus is on mathematical concepts and reasoning. These include randomness, correlation, and probability distributions. But debate continues about whether or not to remove traditional algorithms, in particular, fraction operations, from the curriculum.

The 2000 version of the Singapore mathematics curriculum noted that opportunities should be provided for students to:

- consolidate concepts and skills;
- enjoy meaningful learning;
- participate in cooperative work and broaden learning styles;
- bridge the gap between abstract concepts and concrete experiences;
- explore and attempt different approaches to tasks and problems, and hence observe a variety of consequences;
- shift towards tasks and problems which require higher level of competencies. (p. 18)

Wong (2009a) gave examples of these learning opportunities and they covered the cognitive, affective, and self-learning domains.

3 Modes of ICT Use in Education

Thirty five years ago, Taylor (1980) proposed three modes of computer use in education, called *tutor, tutee,* and *tool*. These modes still provide a helpful classification of ICT use. However, in recent years, there is phenomenal spread of social media in daily life, and this suggests the need to add a fourth mode to Taylor's framework, which Wong (2009a) called the *co-learner* mode. These four modes are used to organise the discussion in subsequent sections.

These modes of ICT use are supported by different learning theories (Johansson & Gärdenfors, 2005; Woollard, 2011). The following is a rough correspondence:

- tutor → behaviourism;
- tutee → constructionism;
- tool → constructivism and metacognition; and
- co-learner → social cognition.

Note that every learning theory has variants that might apply to more than one mode of ICT use, but the major links given above is helpful in pinpointing the most influential philosophy underpinning each of these modes.

Li and Ma (2010) used a different classification of ICT use: *tutorial, communication media, exploratory environment,* and *tools*. On the basis of a meta-analysis of 46 studies, they found that these modes produced similar positive effects on mathematics achievement, and the effect was stronger for special needs students compared to general education students, or in constructivist lessons compared to traditional teaching. However, they could not review the effects of ICT on attitudes towards mathematics because of limited data.

4 Tutor Mode: Learn *from* the Computer

Under the tutor mode, the computer behaves as a tutor who delivers the instruction and evaluates student learning. The students can learn *from* the computer programs at their own pace, time, and venue, under what is also called *personalised instruction.* Recently, South Korea and Japan have experimented with robot teachers in the classrooms. This computer-based tutor mode is an extension from earlier attempts by educators, notably B.F. Skinner[38], to replace poor teaching with well-designed programmed instruction delivered by teaching machines.

Numerous tutorial packages have been developed, used, and evaluated for different subjects and types of students. These packages include simple drill and practice, simulation, computer-assisted instruction (CAI), YouTube lessons (e.g., Khan Academy[39]), e-lectures, and integrated learning systems (ILS). Sophisticated tutorial packages include powerful multi-media features and comprehensive analysis of student data. The main theory underpinning this mode is behaviourism. The behaviourist principles used to design tutorial packages are similar to those used in direct instruction (see Chapter 3). Quality tutorials can be designed following the steps below:

a) Use task analysis to break the topic down into a hierarchy of manageable sub-skills. Let students decide for themselves the entry level to the program, that is, which sub-skill to work on first, to encourage them to take responsibility for their learning. In systems that use computer adapted testing, the program will select the entry level based on the student's prior performance, thereby adjusting learning experiences to students of different abilities.
b) For each sub-skill, devise a coherent series of explanations and worked examples, paying attention to the cognitive loads of these tasks. Deliver them using a combination of text, audios, and videos, keeping in mind principles such as coherence, modality, and redundancy (Mayer, 2006b).
c) After working through a short presentation, students answer a few questions, typically in multiple-choice or fill-in-the-blank format. These questions are often randomised so that if the students were to attempt the same segment again, they will answer different questions. The questions may be in game format.

---

[38] http://en.wikipedia.org/wiki/Programmed_instruction

[39] https://www.khanacademy.org/

d) Students' answers are checked by the program and immediate feedback is given. Correct answers are acknowledged with general praise, preferably for effort rather than ability (Dweck, 2006), or rewarded with points in games. Those who give the wrong answer at the first attempt are usually offered additional chances, after some hints have been provided.
e) The program determines what to present next. This is an important feature of personalised instruction.
f) The whole cycle is repeated until the student completes the whole section for the target topic or logs out prematurely.
g) At the end of each session, student responses can be downloaded and summarised.
h) Students should be informed of their performance and further learning activities can be assigned accordingly.

To design quality tutorial packages requires expertise in contents, course design, and the authoring software; many hours of writing; and painstaking effort to test the product on-site and with the target students. These are usually produced by commercial companies or well-funded research projects. They are also expensive.

Teachers can learn to write simple tutorial with drill and practice using packages such as *PowerPoint*[40]. Teacher-made tutorials can be quite effective, especially for low achieving students because the teacher can design materials targeted to their ability and learning needs. Motivation is increased when the teacher personalises the problems by using students' names (provided they agree to this), and some students prefer watching video explanations given by their teacher to those by strangers.

These tutorials are now used under *blended learning* in *flipped* classroom. Students learn new contents from the tutorials at home, and the teacher uses class time to answer their queries, check homework, or conduct group activities. These hybrid arrangements enable teachers to plan their lessons with greater flexibility. A recent US report noted that

---

[40] http://office.microsoft.com/en-us/templates/multiple-choice-test-4-answer-TC001018386.aspx

students in hybrid classes, including mathematics, outperformed their peers in traditional classes (Schaffhauser, 2014). A meta-analysis (Means, Toyama, Murphy, Bakia, & Jones, 2010) also found a similar, modest benefit covering mostly non-mathematics subjects (only one study was about mathematics), but the authors believed that this effect could be due to the additional learning time given to the blended conditions and not the media per se.

Research into different forms of the tutor mode has produced mixed findings. According to Enyedy (2014), there was little evidence of effectiveness of personalised instruction for different subjects, largely because of "the incredible diversity of systems that are lumped together under the label of Personalized Instruction" (p. i). On mathematics education, the US National Mathematics Advisory Panel (2008) noted that "technology-based drill and practice and tutorials can improve student performance in specific areas of mathematics" (p. xxiii) but "the available research is insufficient for identifying the factors that influence the effectiveness of instructional software under conventional circumstances" (p. xxiv). In a large scale study, Dynarski et al. (2007) found that tutorial products for Grade 6 mathematics and Algebra did not produce statistically significant results compared with control classes, with effect size in the range -0.10 to 0.10. There is some evidence that tutorial packages are beneficial to low achieving students because they can repeat the tutorials as many times as they wish and practise the skills on randomly generated questions. Students are found to express initial enthusiasm for these lessons because of novelty and the welcome breaks from monotonous lessons. But many of them sooner or later prefer traditional lessons for various reasons, such as traditional lessons are familiar, take less time, and allow students to ask questions immediately when they do not understand some parts of the lessons. However, it is difficult to find published reports to support these observations.

5 Tutee Mode: Learn *through* Programming the Computer

Tutee is the opposite of *tutor*. In the tutee mode, students use a programming language to *teach* the computer to carry out certain procedures. Although students can instruct the computer to do so using application software such as word processor, spreadsheet, and dynamic geometry, these are normally considered under the *tool* mode to be discussed later. To be under the *tutee* mode, programming is the essential feature, that is, students learn *through* programming the computer. The underlying theory for this mode is *constructionism* or *learning by making:* students construct new knowledge and develop creativity by making or creating new tangible objects online rather than physical ones, such as animations, games, videos, or microworlds. This theory is based on Piaget's genetic epistemology (1973) and is widely promoted by Hoyles and Noss (1992), Papert (1980, 1993), and other enthusiastic educators. This also aligns with the *maker movement*[41], a contemporary sub-culture focussing on technology-extended DIY (do-it-yourself) approach to learning and applications of the learning.

5.1 *To teach is to learn twice*

The French essayist, Joseph Joubert (1754-1824), noted that "to teach is to learn twice,"[42] so one benefit of the tutee mode is that students can deepen their content knowledge by teaching it to the computer in a language that it *understands.* They break down a sub-topic into a sequence of clearly defined steps and write precise, unambiguous instructions for each step using the vocabulary of the chosen programming language. These steps are compiled into procedures to carry out different tasks, for examples, to compute arithmetic operations, play simple games, or draw graphics on the screen. Doing so strengthens students' mastery of the content, encouraging them to seek out new knowledge, and helping them become more metacognitive by reflecting on their own thinking, learning, and teaching it to the computer.

When programming was first introduced to Singapore schools in the early 1980s, the computer language used was *BASIC* (Beginner's All-purpose Symbolic Instruction Code). This was taught in extra-curricular courses to promote computer appreciation among secondary school

---

[41] http://en.wikipedia.org/wiki/Maker_culture

[42] http://www.brainyquote.com/quotes/quotes/j/josephjoub108036.html

students, but there was no attempt to include BASIC programming in mathematics lessons, which was the case in some countries such as UK (some old mathematics textbooks had a section on BASIC programming). However, BASIC was eventually phased out because of several factors.

- Much learning and effort is required to produce simple messages and calculations in BASIC. Most students found these activities tedious and uninspiring.
- BASIC tends to lead to poor programming practices, derided as *spaghetti code,* where the program is unstructured with many layers of GOTO statements. This does not promote logical thinking.
- New application software appeared in late 1980s. They can be easily learned and used to produce exciting outputs, such as professional formatting and embedding diagrams into texts, which cannot be readily done using elementary BASIC programming.
- The goals for computer in education were changed from computer appreciation to digital literacy, so that students can use application software in lessons and daily life. Programming in BASIC or other languages is not part of this literacy, until it is revived in the past two years; see Section 5.4.

These factors led to the demise of BASIC programming in Singapore schools since the 1990s. However, some countries that still include programming in the mathematics curriculum have switched from BASIC to Logo. Let me begin with a short primer of Logo.

5.2 *A short Logo primer*

Logo is a powerful high-level, yet child-friendly, programming language developed by Papert and his team at the Massachusetts Institute of Technology. The main purpose for teaching Logo is to help students engage in "deepest ideas from science, from mathematics, and from the art of intellectual model building" (Paper, 1980, p. 5). For mathematics, the big ideas are from geometry, numbers, and algebra, and the scientific and modelling processes include forming hypotheses, testing them, and appreciating that making errors is a natural part of programming and learning. Students also develop algorithmic thinking through programming. This section provides a very short introduction to Logo, and the pedagogical implications are discussed in the next section.

Several versions of Logo are available as freeware, and the *MSWLogo* (Microsoft Windows Logo)[43] is used here. There are many useful online resources and books on Logo programming and activities, dating back to the years when Logo has very strong following (e.g., Harvey, 1987; Martin, Paulsen, & Prata, 1985).

On opening the Logo screen in MSWLogo, a small triangle appears as shown in Figure 6.1(a). It represents the *turtle,* which is the key metaphor for Logo as the turtle graphics. The top vertex represents the heading of the turtle, and students have to keep track of the heading during execution. Figure 6.2(b) shows a cardboard cut-out of a turtle, which young students can use to monitor the turtle's motions, adding a kinaesthetic dimension to the programming to aid visualisation of the motions.

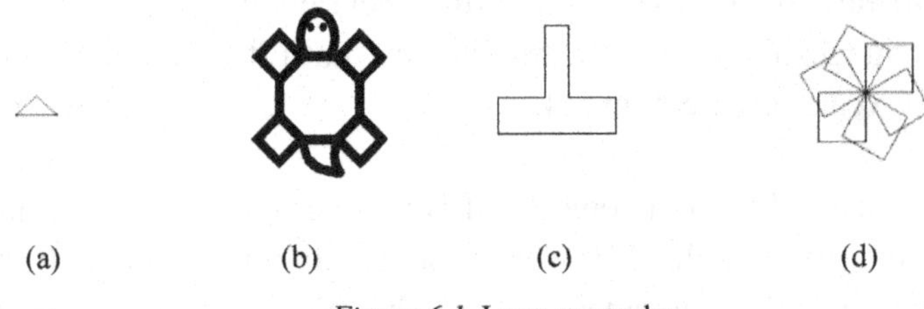

(a) (b) (c) (d)

Figure 6.1. Logo examples

1. *Basic Logo primitives.* Students first learn the five basic primitives: **fd** (forward), **bk** (back), **rt** (right turn), **lt** (left turn), and **repeat** (to repeat a sequence of steps). Using just these few primitives, they can easily draw patterns, such as the one shown in Figure 6.1(c). The procedure **repeat 4[fd 50 rt 90]** will draw a square. This is too fast for students to follow the how the individual steps are executed; to slow down the motion, use the **wait** primitive in between each step, i.e., **repeat 4[fd 50 wait 10 rt 90]**. The **wait** component can be removed when students focus on the

[43] http://www.softronix.com/logo.html

mathematics. They can apply their knowledge of exterior angles to draw regular polygons using similar procedures. In this way, geometry knowledge becomes dynamic.

2. *New procedures.* It is not efficient to repeat the sequence of individual steps if one intends to draw squares of different sizes. To do so, create a new procedure with a variable to stand for the length of its side. Students can decide on the names for the new procedure and variable, hence giving them a sense of ownership. In the following **square** procedure, **:side** is the name of the variable, and its meaning is intuitively much clearer than the algebraic way: *let x be the length of the side of the square.* This intuitive approach permeates all aspects of Logo programming, and this partly explains why Logo is a favourite language for mathematics instruction.

```
to square :side
repeat 4 [fd :side rt 90]
end
```

The **:square** procedure can be modified to define new procedures for other regular polygons, but this requires two variables, one for the interior angle and one for the side. Thus, the number of variables is not limited to only one as in typical algebra lessons in lower secondary grades. Students can do a *turtle walk* by actually walking on the floor or moving the cardboard turtle on the screen to identify the steps for creating the polygon procedures and to mimic these steps using Logo primitives. Some students may make mistakes in calculating the required angles and sides and end up with unexpected shapes. They should be encouraged to think through why that has happened instead of being anxious about trying to avoid errors. Indeed, the "praise mistakes" technique recommended by Posamentier and Jaye (2006) is likely to work well when students engage in Logo projects.

Students can use their new procedures together with Logo primitives to make more complex designs, and this serves to fire up their imagination. For example, ask them to predict what is produced by this sequence: **repeat 6 [square 40 lt 60]** and then check their prediction; see Figure 6.1(d). This guess-and-check activity helps to develop visualisation, linking symbols to pictures on the screen. When they learn for themselves what works or does not work, without the need for correction by the teacher, they become self-regulated learners.

Negative numbers can be introduced geometrically through discovery; ask students to enter examples such as **fd -40**, **square -40**, and **square -40+60**, and describe what they notice. Pair work is effective here.

3. *Integrate programming and writing.* Design a map of objects around the student's environment, similar to the example shown in Figure 6.2(a). Students write down the Logo steps to move around these objects and write a story about their journey. The example in Figure 6.2(b) was done by a Grade 4 student in Brunei Darussalam (Wong, 2004).

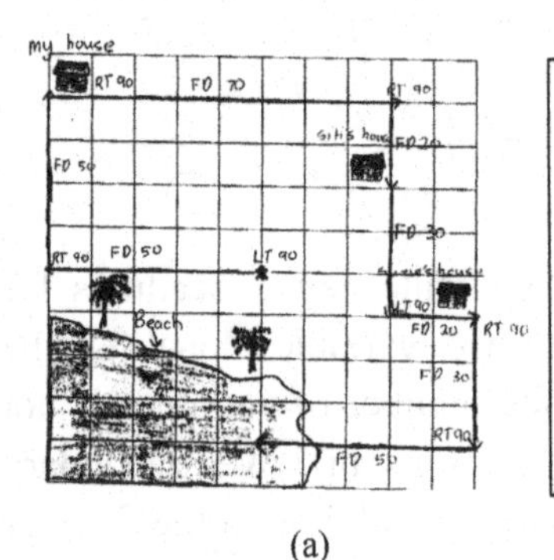

(a)

Last week, a turtle visited my kampung. First it visited my house. I welcomed it. To go to my house the turtle took LT 90, FD 50, RT 90, FD 50. And then the turtle and I went to Siti's house. And Siti welcomed us. To go to Siti's house we took RT 90, FD 70, RT 90, Fd 20. She was very happy to meet the turtle. ...

(b)

Figure 6.2. A Logo journey

4. *Logo microworlds.* A *microworld* is a fairly self-contained environment consisting of a set of pre-defined procedures, and students use these procedures and the Logo primitives to explore specific mathematical topics. However, they do not need to create these pre-defined procedures themselves because they are usually quite complex. Examples are the *numberline* microworld (Oldknow & Taylor, 2000), *ratioworld* (Noss & Hoyles, 1996), *graph microworld* (Wong, 1990a), and *geometric transformation* microworld (Wong, 2001a).

Take the example in Figure 6.3. Students enter **reflect 1 0 new rotate 0 0 90** to determine the image of the object after a reflection in the line

$y = x$, followed by an anticlockwise rotation about the origin. The procedures **reflect, new,** and **rotate** are pre-defined, so that the students focus on the mathematical properties rather than programming, in this case, the effects of combination of transformations.

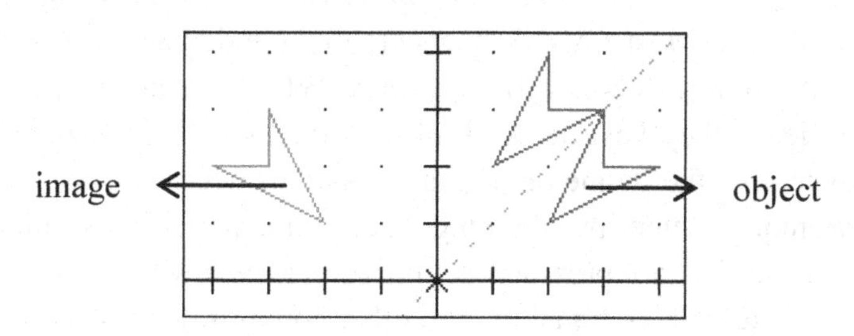

Figure 6.3. Combination of geometric transformations

In an experimental study of Grade 10 students in Singapore ($n = 70$), Ho (1997) found the class that learned transformation geometry using the above microworld did significantly better than the expository class.

5.3 *The Logo approach*

Support for incorporating Logo into mathematics learning can be found in documents by the Australian Education Council (1990), National Council of Teachers of Mathematics (2000), and Royal Society and Joint Mathematical Council (2001), among others. Advocates of Logo have developed numerous Logo activities for classroom use (Clements & Battista, 2001; Ernest, 1989; Fletcher, Milner, & Watson, 1990; Harvey, 1987; Lew & Jang, 2012; Neyland, 1994; Oldknow & Taylor, 2000; Prestage & Perks, 2001). Even though research findings are inconclusive, Logo programming can purportedly bring about the following worthwhile learning objectives.

a) *To master mathematical concepts and skills.* As students manipulate the Logo turtle on the screen, they deepen their understanding of geometric properties such as angles, orientations, symmetry, and geometric transformations. To design complicated shapes, they apply algebraic concepts and rules. Logo's immediate responses to students' input help them better appreciate the link between geometry and algebra. Where available, students can control floor turtles via the procedures they have created, and this kinaesthetic mode is especially helpful for young children who just begin to learn about programming.

b) *To develop mathematical processes and problem solving strategies*. Most Logo projects are open-ended and can be tackled at different levels in different ways. When students work on these projects, they have to think of hypotheses, test them by carefully observing the online output, and make the necessary changes. This promotes scientific thinking and mathematical reasoning. The research has not provided definitive evidence whether or not these strategies can be transferred to other contexts, but this is worthy of support.

c) *To appreciate the benefits of errors.* In the Logo philosophy, errors are treated as either *bugs* which can be fixed or unexpected results that call for explanations of why they happen. Hence, students tend to be less anxious about programming errors, to become confident in handling them, and to develop positive self-esteem.

d) *To promote geometrical imagination.* Students can create striking geometric designs using more advanced features of Logo, including 3D features in the newer Logo versions. This promotes creativity and imagination in geometry, which is lacking in current mathematics lessons, where the main tasks are about calculating angles and sides or proving fairly obvious results. Logo tasks can be integrated with writing, as shown earlier, to add another dimension of creativity.

e) *To develop new learning skills*. Instead of working independently, students are encouraged to complete Logo projects in pairs or small groups. They develop cooperative learning skills, learn to discuss mathematics, and are more readily to engage in this type of learning (Hawkridge, Jaworski, & McMahon, 1990). Kramarski and Mevarech (1997) showed that Logo combined with metacognitive training led to better graph construction compared to Logo without the metacognition.

f) *To acquire programming skills*. This is achieved by learning Logo primitives and creating new procedures. Designing new procedures based on Logo primitives and previously defined ones parallels the way mathematicians create new mathematics through a hierarchy of concepts and skills. Traditional mathematics learning does not provide this valuable mathematisation experience, and Logo programming can help to close this gap.

g) *To enhance motivation to learn mathematics*. By personalising the names for new procedures and variables, students gain ownership of their work. This is a powerful intrinsic motivator. They can immediately see on the screen how changing parts of a Logo procedure leads to surprising shapes and patterns. This works like a fun game. One hopes that this experience will enhance students' attitudes towards mathematics (Ernest, 1989).

5.4 *Trends in tutee mode*

While Logo is widely acknowledged as having positive effects on mathematics learning in the West and is still included in mathematics lessons in some countries, its uptake in Southeast Asia has been weak. During the first wave of introducing computers into schools in the 1980s, Logo was taught in computer clubs in some schools in Indonesia, Malaysia, and Singapore (Hawkridge, Jaworski, & McMahon, 1990). In a survey conducted on 37 secondary schools in Singapore, only nine schools taught Logo (Wong, Lim, & Loh, 1989). There was no integration of Logo programming with school mathematics. The second wave in the 1990s saw the increasing use of multi-media tutorial programs designed to teach specific mathematics content under the tutor mode. This was followed by the tool mode of using application software in mathematics lessons, inspired by constructivism. Given these changes, programming virtually disappears from the Singapore school curriculum. Hence, many Singapore mathematics teachers are not knowledgeable in Logo or BASIC and they do not appreciate the learning objectives of programming mentioned above.

In the past few years, however, there is a revival of interest among some countries, including Singapore[44], about teaching programming in the main school curriculum or as extra-curricular activities. It is believed that learning programming will equip students with the skills and motivation to become designers of digital products in the future to bring economic values to the students and the country. An ambitious reform[45] in England requires all students aged 5 to 16 to learn computer programming as a core subject, starting from September 2014. This course covers debugging, algorithms, computer networking, and other ICT skills. No specific programming language is recommended, but some UK schools are teaching Scratch[46].

Mathematics teachers in Singapore must keep abreast of this trend, become more conversant in this neglected aspect of their training, and make their mathematics lessons part of this initiative.

6 Tool Mode: Learn *with* the Computer

Many software programs developed initially for practical use in business, writing, architecture, government, industry, science, statistical analysis, mathematics, and other fields have been modified for classroom learning under the *tool* mode. The popular ones include the following:

- Spreadsheets and data handling software for numerical computations and graphing, e.g., *Excel*, *BrightStat*; these tools are especially helpful for students who are weak in manual computations, so that their poor technical skills do not prevent them from exploring new mathematical concepts.
- Graphing programs for easy plotting, e.g., *Graphmatica*.

---

[44] http://www.moe.gov.sg/education/programmes/gifted-education-programme/enrichment-activities/secondary/computer-programming-course/

[45] https://www.gov.uk/government/publications/national-curriculum-in-england-computing-programmes-of-study/national-curriculum-in-england-computing-programmes-of-study

[46] http://scratch.mit.edu/

- Dynamic geometry for studying geometry objects and properties interactively. Linkages between geometry and algebra can be explored, e.g., *GeoGebra, Geometer's Sketchpad, Google Sketchup* (3D sketches, which can be printed on 3D printer).
- Computer algebra system (CAS) for solving problems in symbolic algebraic mode, e.g., *Mathematica, Maple, Matlab*.
- Problem solvers for obtaining answers to problems, including step-by-step solutions in some cases. It is a useful aid for problem solving, but students may try to avoid doing their homework by submitting answers they obtain from these websites, e.g., *WolframAlpha, Math Warehouse*.
- Applets for simulating mathematical processes and teaching strategies, e.g., *SimCalc*[47].
- Word processors and presentation tools for students to prepare and present reports of their projects, e.g., *Word, PowerPoint, OpenOffice*.

Some of the above programs are very versatile and can handle more than one type of mathematical procedures. This capability enables teachers to design activities across different topics. Examples of these tool-mode activities can be found in numerous websites and Wong (1997b, 1998a, 2000), and some examples are elaborated below.

This tool mode is supported by theories of discovery learning (Bruner), social constructivism (Vygotsky), multi-modal strategy, and metacognition. Well-designed lesson plans to embed tool features will promote conceptual understanding and reasoning.

6.1 *Stages of use*

There are three stages of using materials in the tool mode.

Stage 1. Use pre-designed templates

Templates are pre-designed so that students can discover mathematical properties by changing some given parameters and trying

[47] http://math.sri.com/

to understand the ensuing effects on their own, without having to refer to the teacher for feedback all the time. This works like Logo microworlds. Students treat these templates as *black boxes* and focus their attention on the functions of the parameters. Templates of different topics can be downloaded from the Internet and used with only minor changes. The two examples given below show these templates are designed to serve different learning objectives.

Example 1. Decimal representation of rational number. This is discussed in Chapter 3, Section 2.3. The *Excel* template in Figure 6.4 can be easily created to generate decimal representation of any rational number. Part (a) shows part of the decimal; in this case, 142857 is the recurring part; part (b) shows the formula to be entered in the cells in rows 4 to 6, based on the standard division algorithm. Students do not need to enter these formulae so that they can investigate ideas such as the length of the repetend.

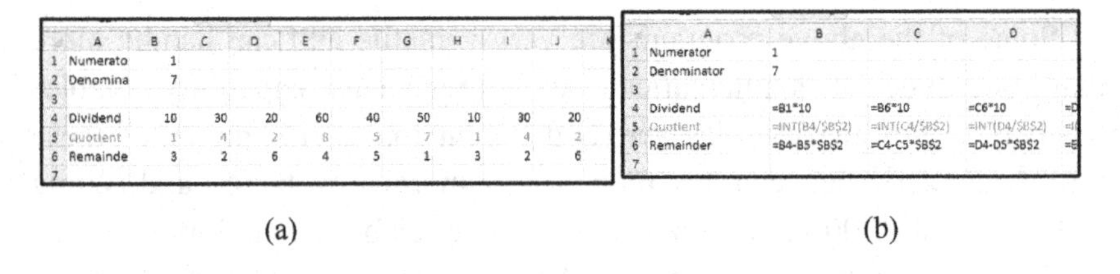

	A	B	C	D	E	F	G	H	I	J
1	Numerato	1								
2	Denomina	7								
3										
4	Dividend	10	30	20	60	40	50	10	30	20
5	Quotient	1	4	2	8	5	7	1	4	2
6	Remainde	3	2	6	4	5	1	3	2	6
7										

(a)

	A	B	C	D
1	Numerator	1		
2	Denominator	7		
3				
4	Dividend	=B1*10	=B6*10	=C6*10
5	Quotient	=INT(B4/$B$2)	=INT(C4/$B$2)	=INT(D4/$B$2)
6	Remainder	=B4-B5*$B$2	=C4-C5*$B$2	=D4-D5*$B$2
7				

(b)

Figure 6.4. Excel template for decimal representation of rational number

Example 2. Mean as even distribution. Students frequently learn *mean* as the result of dividing the sum of the given numbers by the number of items (or groups). They rarely appreciate the following properties:

a) finding the mean is an attempt to divide the numbers evenly into the given number of items (groups);
b) the mean lies between the minimum and maximum values given;
c) the sum can be obtained by multiplying the mean by the number of items; and
d) the mean is affected by extreme value(s) in the given numbers.

To help students acquire these conceptual ideas, the *Excel* template in Figure 6.5 can be easily created. The bar chart illustrates ideas (a) and (b); idea (c) is shown by the results in cell B12; to explore idea (d), students can enter one extreme value in a cell from B5 to B9 and notice what happens. Additional rows can be added to increase the number of items or groups for further investigation.

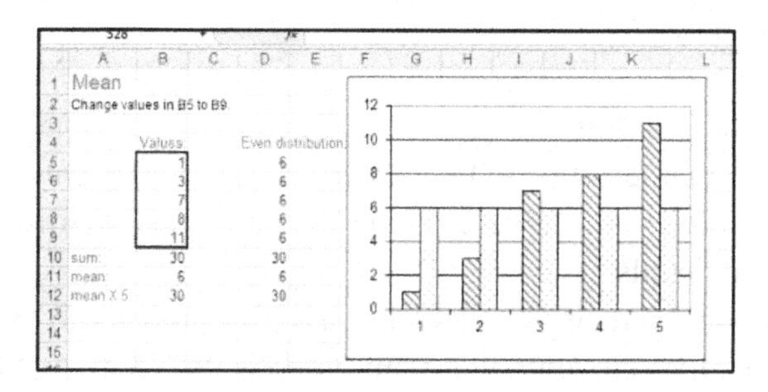

Figure 6.5. Excel template for exploration of *mean*

Stage 2. Create simple templates

At this stage, students learn to create simple templates on their own. For *Excel*, this might involve entering formulae to carry out computations and plotting various types of charts. For dynamic geometry, students can create standard geometric objects according to their defining properties so that the objects are *drag invariant*, i.e., these properties (e.g., four right angles of a rectangle or two equal sides of an isosceles triangle) remain unchanged when the object is dragged on the screen. This stage is similar to programming in the tutee mode.

Stage 3. Become advanced users

At this advanced stage, students continue to learn more powerful features of the software, such as creating macros using Visual Basic in *Excel* or sliders and scripts in *Geometer's Sketchpad*. Then they investigate complex mathematics using these features, similar to the way mathematicians discover patterns using computer as a research tool. However, few students reach this stage due to lack of curriculum time and motivation, different goals for the mathematics curriculum, or weak guidance from their teachers. The enthusiastic teachers should strive to reach this stage themselves.

6.2 *Teaching sequence*

The suggested sequence of using pre-designed templates is: off-line introduction → computer explorations → off-line reinforcement. This is elaborated below.

1. *Off-line introduction*

Introduce the mathematics to be learned so that the students have an idea of what to focus on. However, it is not necessary to disclose all the mathematical details.

On the one hand, it is relatively easy for students to infer specific *properties* such as an angle in a semi-circle is 90° from a table of values generated by using dynamic geometry. On the other hand, it is harder for students (even teachers) to infer *new concept* from online output. Without knowledge of the target concept, students tend to pay attention to only superficial features rather than to think deeply about the mathematics. This was mentioned as the *to know is to see* phenomenon in Chapter 4, Section 3.1. Two examples are used to illustrate this.

First, consider Figure 6.5. Many students notice that five bars of the mean are plotted, but they may not be able to arrive at the concept of *even distribution* by just looking at the graph, if they have not been taught this concept earlier.

Second, consider Figure 6.6. Part (a) shows the quadratic graph and the absolute-valued graph plotted in the typical range of values using *Graphmatica.* When these two graphs are zoomed in repeatedly around the origin, the shape of the absolute-valued graph remains unchanged, but that of the quadratic graph becomes almost linear, as shown in part (b). What is the underlying mathematical property that can be observed from this exploration?

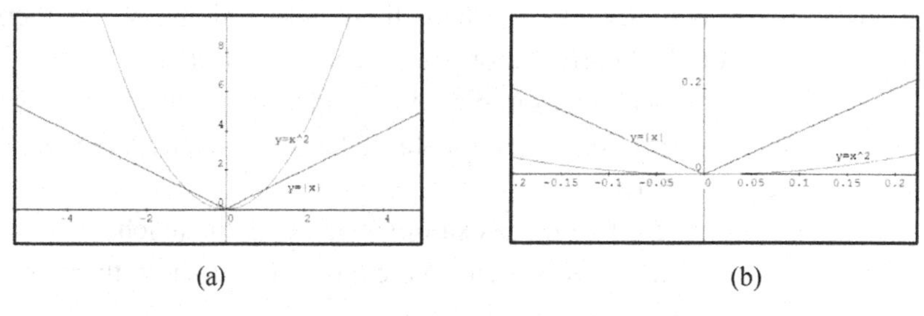

(a) (b)

Figure 6.6. Local straightness and differentiability

Almost all the teachers who were shown this in my courses noted these changes but none of them were able to *discover* that this shows the concept of differentiability in terms of local straightness (Tall, 1989), because this concept was new to them. Hence, teachers who intend to use a similar approach to let students discover new concepts rather than properties should carefully plan this off-line introduction so that curriculum time is not wasted on trivial work.

2. *Computer explorations*

This can be conducted in four different ways, progressing from teacher-centred demonstration to student-centred open investigation.

a) Whole class demonstration. The teacher projects the template onto a big screen or interactive whiteboard so that it can be seen by all the students. The teacher shows several examples and counter-examples on the template and explains the mathematics as clearly as possible. The template should enable the teacher to show as many examples as needed, varying the mathematical and perceptual features as recommended by Dienes (1964).
b) Whole class discussion. This engages students to think actively about the activity. Train them to ask *what if* questions and use the template to check their answers. A typical sequence of teacher moves is as follows:
   i) Show the first example and ask students to write down two or three observations. Make sure that they can see the projection clearly.

ii) Ask a few students to share their observations. Write them down on the board without comments at this stage.
iii) Show the second example. Ask students to add at least one more observation or eliminate at least one observation from their initial list.
iv) Repeat with a few more examples given by the teacher.
v) Ask them to suggest their own examples and show them with the template.
vi) If they decide that they have enough examples, discuss the observations to arrive at some mathematically significant conclusions.
vii) Challenge them to justify the conclusions.
viii) Summarise the findings.

c) Hands-on activity with worksheet. This should be conducted in a computer laboratory. Provide students with the worksheet so that they can work individually, in pairs, or in small groups. Ask them to use the calculator to check the online output or to sketch the online graphs or diagrams onto paper; this stimulates active learning through *tinkering*. The worksheet should suggest checking typical as well as limiting cases. During the lesson, walk around to check that the students are on task, and give help whenever necessary. Stop the class periodically to discuss or summarise findings. Ask students to share any special observations they have discovered. Collect their worksheets and mark them as class work.
d) Open investigation. Only the template is given and students make their own conjectures and check them by working through the template on their own. To reach this stage, students must be trained through the earlier stages.

3. *Off-line reinforcement*
Once the computer activity has been completed, give further practice to reinforce the concept and develop the skill to the required standard.

In a survey of secondary mathematics teachers in Singapore, Leong (2003) found that their most preferred use of Geometer's Sketchpad was distributed as follows: teacher demonstrations (20); use of software to prepare worksheets and tests (7); student hands-on (3). This suggests that the inquiry approach was not popular about a decade ago, but it is not clear whether more teachers are now conducting inquiry lessons using the tool mode.

6.3 *Teaching and learning issues about tool mode*

This section deals with several issues when ICT is used as a tool mode in mathematics lessons. These issues need to be fully addressed before the power of the tool mode to promote higher order mathematical processes can be realised.

First, teachers need to note that some of these tools have limited precision due to limitations on the pixels on the screen and the software capability. Graphs with large parameters may not be plotted correctly. Numbers are stored up to a fixed number of significant figures, such that the computed values are not exact; see Figure 6.4. Students should also be made aware of these limitations.

Second, consider the effect of the *knowing is seeing* phenomenon. When students submit work that contains only superficial conclusions, especially when new complex concepts are involved, teachers need to give meaningful feedback to facilitate further learning. Consider the conclusion given in Figure 6.7. What would be some helpful feedback to give?

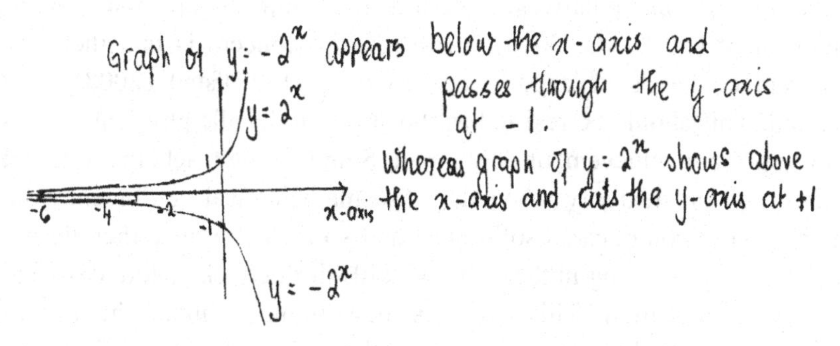

Figure 6.7. Superficial conclusion from graph of exponential function

Third, the online output may be interpreted by the students in ways different from the mathematically correct one. A commonly used task is to investigate the role of c in the linear equation, $y = mx + c$. After plotting $y = x$ and $y = x + 3$, say, many students infer that the second graph is transformed diagonally as shown in Figure 6.8(a) because it looks quite *natural,* whereas the mathematical interpretation shown in Figure 6.8(b) is not obvious and requires further thinking. It seems that online images are more persuasive than symbolic argument, and this can mislead students into making the wrong inference. Careful marking of student work can help to spot such misinterpretations.

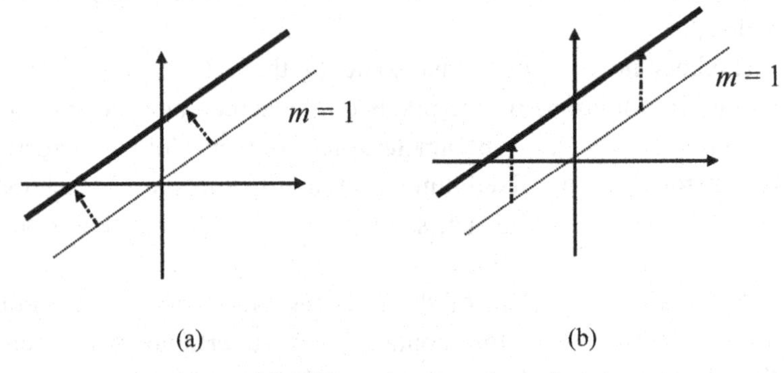

(a) (b)

Figure 6.8. Different interpretations of y-intercept

Fourth, some of these tools can be used to create virtual manipulatives, which are usually more accurate replicas of the real ones. These virtual manipulatives have positive impacts on mathematics achievement (Li & Ma, 2010). However, the concern is that their use may replace hands-on experiences, as noted by Ofsted (2008) from England. This should be resisted as the ability to handle physical objects is an important achievement on its own. Some young teachers are found to lack manual dexterity when they use mathematical equipment, fold paper, cut out complicated solid nets, and so forth, because they do not receive enough training in these motor skills during their school days and pre-service education. This calls for new thinking about the use of concrete vs. virtual manipulatives in mathematics instruction (Clements & McMillen, 1999).

Fifth, *PowerPoint* is now a ubiquitous tool in lessons. It is useful when intricate diagrams or animations are shown to the class. However, its use to teach mathematics has several drawbacks, which teachers should be aware of (cf. Tufte, 2003):

- Disclosing mathematical solutions line-by-line in *PowerPoint* slides tends to reduce meaningful question-and-answer interaction and discussion between teacher and students, because there is a strong tendency to follow the pre-determined sequence of the slides rather than deviating from it to spontaneously follow up responses from the students.
- It is difficult to demonstrate the problem solving process, as the teacher talks aloud, moving from one step to the next without being constrained by the slides.
- Many teachers read the slides with their back facing the class, and this quickly results in classroom management problems.
- At the end of a sub-topic, a well-planned presentation on the board (say, divided into columns) will capture the flow of the main ideas and allow for cross referencing. This coherent presentation and flexibility cannot be achieved using *PowerPoint* slides because the slides are transient and it is not possible to project all the slides at once on the screen. When this tool is used frequently, some teachers have lost the important skill of coherent board presentation.

Finally, students with smartphone or tablets may search for answers online rather than think through the mathematics. For example, when teachers in my in-service classes were asked to think about "why 1 is neither a prime nor a composite number," someone would immediately read out an answer from a Google search, instead of reasoning it out. Rather than banning the use of these tools in my lessons, I challenged the teacher to explain the looked-up answer in his/her own words. This will work with students as well.

To summarise, this long list of issues does not imply that the tool mode is too complex or problematic for classroom use. The power of tool use is well recognised, and taking steps to overcome some of the issues above can change the inherent power into strong student achievement.

7 Co-Learner Mode

This new mode highlights the roles of cooperation and networking in using ICT to promote interdependent teaching and learning. Under this mode, teachers and students learn from one another when they construct knowledge by working through shared information of mathematics found from many Internet resources. They can use social media, blogging, discussion forum, and other communication media to communicate and share their learning. International collaboration can be involved in this co-learner mode, such as the GlobalSchoolNet.org[48].

This mode arises from the growing acceptance of social-constructivist learning and a proliferation of professional learning communities (PLC) in education. Under PCL, there is no explicit and sole authority of new knowledge. This also tempers the traditional authoritative role of the teachers as the main transmitter of knowledge. This can result in more cordial teacher-student relationships as well. In fact, there is a Chinese proverb which says that the students must excel their teachers in learning for it to be judged as successful.

The first area of co-learning is about mathematics topics that are new to both teachers and students. This can arise in several ways:

- The contents are additions to the mathematics curriculum for which the teachers have not received any training.
- Students have posed interesting mathematics questions that cannot be solved using what have been learned.
- They are applications and modelling scenarios that require new knowledge and data.

Besides the acquisition of new knowledge, the co-learner mode also involves teachers modelling the learning strategies they have engaged in to acquire new knowledge. They show by personal practice that looking

[48] http://www.gsh.org/GSH/pr/index.cfm

up information is only the first step of knowledge construction. This should be followed by relating the new information to previous knowledge and experiences, comparing and contrasting alternative explanations found online, subjecting them to mathematical reasoning, and seeking further assistance to clarify difficult points. Through this co-learner mode, teachers who continue to learn will appreciate the difficulties encountered in learning and the joy of overcoming these challenges. Thus, they will be more empathetic about the learning problems faced by their students. ICT tools can facilitate this co-learning to a level not feasible in the past. Since this is a very recent development, more practical resources and investigations are needed.

8 Computer-based Automatic Assessment (CAA)

This section is written specially for education administrators, education researchers, and instructional designers. Classroom teachers may not have the competency and time to create such a system for their own classes, unless they also become researchers of their own practice or complete a similar project as part of their certification for higher degrees.

Administrators of public examinations in many countries are interested in designing and implementing these summative examinations using computer-based automatic assessment system to save cost, to improve efficiency, to shorten time between examination and release of results, and to ensure reliability of testing. Recent examples of this approach are the two consortia in US called the Partnership for Assessment of Readiness for College and Careers (*PARCC*) and Smarter Balanced Assessment Consortium (*SBAC*). Education researchers and instructional designers are interested in using this approach to study students' learning by capturing comprehensive data about their performance. An important feature of these systems is a fully automatic grading system.

For mathematics testing, automatic grading can be easily implemented for selected response items (e.g., multiple-choice) and simple fill-in-the-blank items, but there are considerable technical challenges to automatically grade construct-response items, which requires students to supply their own solutions. The difficulties include checking that the many possible solutions submitted by the students are mathematically correct, giving partial credits to incomplete solutions, and handling the symbolic and graphing representations special to mathematics. These technical challenges have not been fully resolved.

It is difficult to find CAA systems that provide feedback to the students about their solutions, other than a grade after their performance. From the perspective of pedagogy and formative assessment, a fully automatic and feedback-less system does not facilitate future learning. An alternative to CAA is to design a system that provides only semi-automatic grading but customisable feedback for the students, using principles of formative assessment (e.g., Hattie, 2009; Wiliam, 2011). The Singapore Mathematics Assessment and Pedagogy Project (*SMAPP*) has produced a low-cost but working prototype to show how these features can be implemented (Wong et al., 2012). This prototype is shown in the schematic diagram in Figure 6.9.

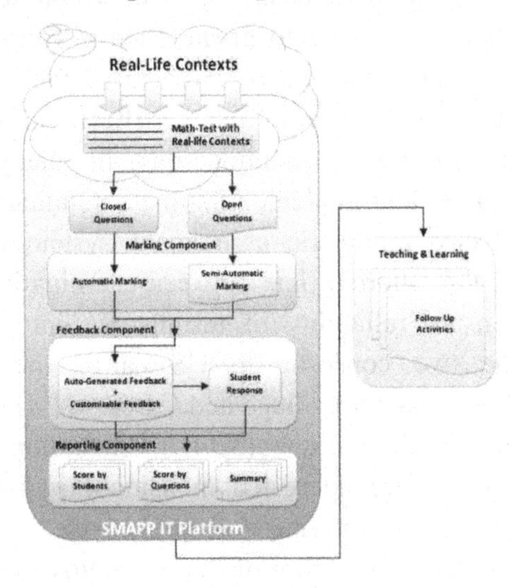

Figure 6.9. Key components of the SMAPP IT prototype

The SMAPP IT prototype consists of the following seven components and the rationale for each component is explained below (Wong et al., 2012).

1) *Delivery.* The system delivers mathematics extended tasks with real-life contexts. Simple animations are also included.
2) *Data capture.* Students work on the tasks and answer the questions. A closed question requires them to choose from multiple choice options or to enter the final answers into designated boxes. An open question requires students to type in their mathematical workings and answers into text boxes. A text editor allows them to enter mathematical symbols by clicking on an equation tool bar. An online calculator is also provided. Besides mathematical problems, some open questions ask for student opinions, such as the benefits of paper cycling, which do not have right or wrong answers, but are included to assess their ability to communicate ideas clearly. The system captures these responses online.
3) *Automatic marking of closed questions.* The closed questions are automatically marked. This saves marking time for the teachers and ensures consistency in marking.
4) *Semi-automatic marking of open questions.* Semi-automatic marking is recommended for open-ended questions. Each *short* open question has several anticipated responses, which may be correct, partially correct, or wrong. The system suggests the most likely mark to be assigned to an answer, but the teacher should review this. More *complex* open questions with multi-step workings or text input will not be marked by the system; instead, the system provides a marking scheme from which the teacher can choose the most appropriate mark. An example is shown in Figure 6.10.

Under this semi-automatic marking of open questions, teachers need to pay due attention to student thinking through checking their solutions, which is an essential step in formative assessment. The marking scheme also provides some consistency in grading.

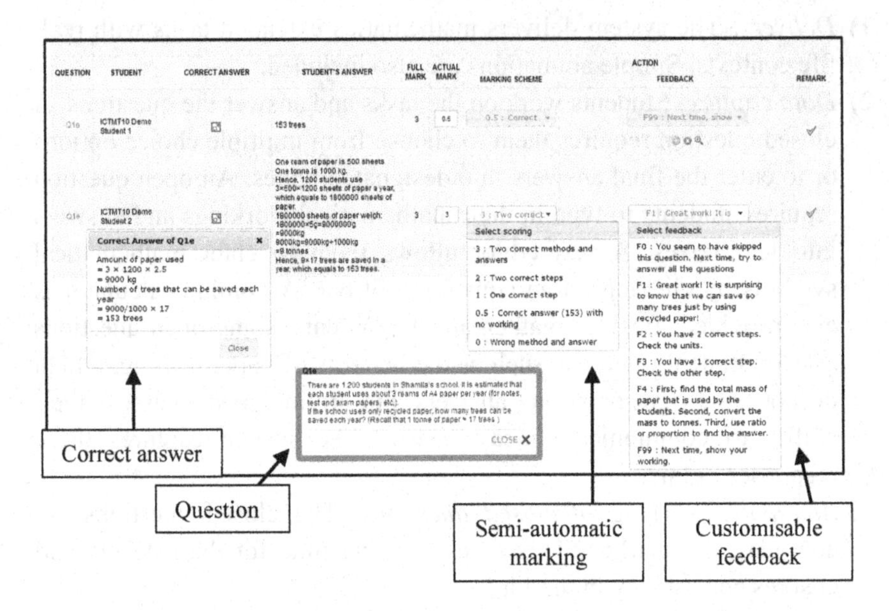

Figure 6.10. Marking and feedback components: Teacher version

5) *Auto-generated and customisable feedback.* Every question is tagged with several feedback comments. For correct answers, the feedback includes giving general praise, reiterating the key steps in case the students get the correct answers by the wrong or inefficient methods, or asking students to solve a related problem as an extension.

   For partially correct or wrong answers, either hints or the full solutions are given. If the student gives only the final answer where working is required, the system includes the feedback, "Next time, show your working." Samples of these feedback comments are also shown in Figure 6.10.

   Teachers can add their own feedback comments, which are automatically made available to teachers from the same school. Customizable feedback refers to this situation where the teacher can select from the given feedback comments or add their own in order to match students' solutions.

The teacher can mark the answers by individual students or by questions. This online system is much efficient and more consistent than if the teacher were to write similar feedback comments onto the paper scripts.

6) *Students' responses to online feedback.* When the students receive the feedback comments online, they are required to respond with one of the following three options:
   a) Now, I understand.
   b) I still do not understand, so I will discuss with my teacher.
   c) I still do not understand, so I will discuss with my friends.

   Under normal situations, teachers rarely find out what students think about the feedback written on their homework, worksheets, or test papers. This component is a first attempt to fill in this gap in teacher knowledge. Only brief options are given here in order not to burden students with elaborate choices. Nevertheless, making them aware of this step may promote a basic form of metacognition.
7) *Report system.* The system automatically generates reports by questions, students, and classes. All the data and reports can be downloaded into *Excel* for other analyses. The data are useful as evidence to plan follow-up activities.

Online surveys and interviews with selected Grade 7 students found mixed responses to this system. Some students mentioned the difficulty they had with entering mathematical expressions and using the online calculator. They preferred scribbling rough workings and writing answers on paper to the online system. Some noted the advantages of online assessment, and this is illustrated with the comment below:

> That we can easily search for information online and we can erase our answer easily and quickly, unlike using paper, and we do not need to waste paper.

Efforts should be directed to improve this prototype of online formative assessment system and to equip students with the skills to effectively use similar system to improve learning. This aligns with the international trend called *technology-enhanced assessments* (TEA) (NCTM Review Committee, 2013).

9 ICT Questions

To effectively use digital tools in mathematics learning, students need to go beyond ICT skills and focus on mathematical contents and processes. A few questions to help them become aware of this use of ICT are:

- What changes do you notice when the parameters are changed? What mathematical ideas might explain these changes?
- Do you change the parameters in a systematic rather than an ad hoc way?
- What questions can you ask about the given context? How can the ICT tools help you to find relevant answers?
- How can you use the software to check your answers, without asking your teachers to do it for you all the time?
- How does working on the ICT activity prepare you to solve similar problems manually? (Calculator is allowed.)

10 Concluding Remarks

There are too many areas of ICT use not covered in this chapter; see Hoyles and Lagrange (2010) for research about other recent issues. Nevertheless, what is evident from this brief survey is that there are many innovative ideas and practices of using the four ICT modes in mathematics instruction. But there is not yet unequivocal evidence about the impacts of ICT use, partly because of the small number of rigours studies as identified by meta-analysts. Findings about old technologies and previous generations of students do not shed much insight about the so-called *digital natives* (Prensky, 2001), who are very familiar with social media and sophisticated digital tools. Evidence has shown that the brains of these digital natives may be wired differently due to their early immersion in handheld digital devices. Educators and parents are also concerned about many new digital issues: increasing incidence of digital addiction, depression, ease of distraction, loss of sustained attention to a task, the illusion of parallel processing, instant response, cheating, and so on (e.g., Public Health England, 2013).

While enthusiastic advocates and thoughtful critics continue to disagree on many of these pertinent issues, mathematics teachers may like to choose one ICT mode and tool, become advanced users, experiment its use in their lessons, and work with researchers to assess the impacts. This chapter's epigraph by Bill Gates highlights the important role of teachers rather than technology to motivate students to learn. This concerns the Attitude factor of the intended curriculum, which will be dealt with in the next chapter.

Chapter 7

Attitudes: Energise Learning with Emotional Power

The cognitive-affective division is a popular and powerful discourse about learning objectives and outcomes, but neuroscience shows that cognition and emotion are inextricably intertwined in neural processing. This chapter delves into this division and then examines the meaning, measurement, research, and interpretations of affective constructs, especially attitudes and motivation. The last section presents the *M_Crest* framework, consisting of six different types of motivators to entice students to learn mathematics: *Meaningfulness, Confidence, Relevance, Enjoyment, Social relationships,* and *Targets*.

Willing is not enough; we must do. Johann Wolfgang von Goethe (1749 – 1832)

1 Cognitive Domain vs. Affective Domain

In mid-1950, Bloom and his colleagues (Bloom, 1956) divided learning outcomes into three domains and three corresponding taxonomies:

- Cognitive domain: use mental skills to process information, create knowledge, solve problems, and remember things.
- Affective domain: expressions of emotions, attitudes, and values.
- Psychomotor domain: engage in manual or physical skills.

This seminal work has spread throughout the world and the taxonomies have become the core framework for specifying learning objectives and assessing learning outcomes in these three domains. The cognitive-affective division in their work has become entrenched in the education community, providing a meaningful and powerful language for education discourse about learning objectives and outcomes. It encourages teachers to include both types of learning outcomes in their lessons, instead of only focusing on cognition.

Research in neuroscience [49] in the past two decades, however, challenges this division. It shows that emotions impact on cognitive functions, for example, anxiety can inhibit working memory and joyful experiences can facilitate retention (Willis, 2007). Conversely, cognitive activities can generate different emotions. In very brief terms, incoming sensory data have to pass through emotion regions of the brain, mainly the amygdala and hippocampus, before they are processed cognitively in the prefrontal cortex. This seems to also occur during retrieval of information from long-term memory. Since cognition and emotion are inextricably intertwined in neural processing, the cognitive-affective division should be understood as a useful heuristic device in education discourse rather than a state about reality.

2 Types and Meanings of Affective Constructs

The affective domain is overwhelmed by numerous ill-defined terms or constructs. Different terms may mean the same thing with only minor differences, or the same term may be given very different meanings by different authors. In a similar way, instruments with the same label may or may not measure the same affective constructs. For examples, an instrument purported to measure *attitudes towards mathematics* includes items about feelings towards mathematics teachers and their teaching methods. Hence, users of these instruments must be able to distinguish between labels and contents.

---

[49] http://www.scholarpedia.org/article/Cognition_and_emotion

A few commonly used affective constructs are described below. They have been arranged roughly along the cognitive-emotive continuum proposed by Grootenboer, Lomas, and Ingram (2008), starting from the cognitive end.

- *Beliefs*. What do I *think* it is or should be? These include beliefs about the nature of mathematics and how it should be taught and learned. In the TEDS-M study (Tatto et al., 2012), beliefs about the nature of mathematics are classified under two scales: mathematics as a set of rules and procedures, and mathematics as a process of inquiry. Samples of these two kinds of beliefs are:
    - Mathematics is a collection of rules and procedures that prescribe how to solve a problem. (ibid, p. 154)
    - In mathematics many things can be discovered and tried out by oneself. (ibid, p. 155)

  In general, higher percentages of the future teachers endorsed the inquiry belief than the procedure belief (about 75% vs. 60% for the Singapore sample), but the two sets of beliefs are not incompatible to each other.

  Other beliefs about mathematics can be found in Chapter 1, Section 1.
- *Values*. What are the reasons underlying the choices made? The reasons could be based on thinking and/or feelings. The purported values for studying or teaching mathematics are closely related to the goals of mathematics education; see Chapter 1, Section 3.
- *Personal attribution*. What affect my *actions* and what can I *do* about it? People attribute success or failure differently to effort, ability, luck, task demands, and expectations from significant others. Self-efficacy is one's belief in own capacity to produce a desired effect. The *attribution bias* states that people often attribute success to personal ability and failure to external factors. These attributes can affect students' engagement in their learning.
- *Attitudes*. What is my *favourable* or *unfavourable* expression or position about the attitude object? Common attitude objects include mathematics, learning mathematics, classroom events, and teachers. Attitudes may be considered weaker versions of beliefs

because the latter has a stronger cognitive element, which is more resistant to change compared to feelings associated with attitudes.

- *Emotions or Feelings*. What do I *feel* about this? This terms cover anxiety, confidence, joy, motivation, and so on. The term *affect* is often used in a narrow sense to mean this category.
- *Perceptions* or *Perspectives.* This is the catch-all term used to cover some or all of the above constructs. It emphasises ideas and feelings gained in a subjective rather than an objective way.

Educators should explain the meanings of the affective constructs they write about by using the three different approaches described below.

Lexical definition. This refers to meanings given in common dictionaries, including etymological roots of the terms. For example, the online Dictionary.com defines *confidence* as "belief in oneself and one's powers or abilities." [50] Lexical definitions often lead to circular loops; for instance, looking up *belief* in the above dictionary leads to *confidence.* However, this approach is still helpful if the lexical definitions are made up of basic words whose meanings do not require further definitions.

Conceptual definition. In this case, the writer provides own explanation of the construct by discussing the *essential* ideas, giving a list of attributes, and illustrating the construct with appropriate examples. Conceptual definitions often include words about other constructs that may need to be explained. These conceptual definitions are often given at the beginning of journal articles, in a section called *Definitions of Terms* in dissertations, and in specialised dictionaries, encyclopaedias, and books. These specialised sources can be in print (e.g., Colburn, 2003; Inglis & Aers, 2008; Mason & Johnston-Wilder, 2004; Ravitch, 2007) or online, e.g., *Wikipedia*, *ASCD* (2014).

Operational definition. This describes the operations or procedures used to identify whether a given object or event has the intended properties associated with the construct. The operations often involve administering

---

[50] http://dictionary.reference.com/browse/confidence?s=t

research instruments, counting occurrence of the target behaviours, or rating the behaviours by qualified researchers. The resulting measurements then become *variables* used to represent the constructs. This is the "construct is what the instrument measures" approach, and the most well-known example is, "intelligence is what IQ test measures." This approach is not satisfactory because it averts the necessity to clarify the construct independently of its measurement. The developer of the operational definition should explain why the measurements obtained from the operations are valid and reliable indicators of the intended construct by referring these measurements to a carefully formulated conceptual definition. Unfortunately for education, the same construct has been defined operationally by instruments that differ significantly from one another in the items used and the operations of collecting the data. This ambiguity has resulted in weak theoretical foundation for research into the affective domain, making it quite difficult to summarise the vast number of studies in this field.

As an example of the above approach, the construct, *motivation to learn mathematics,* is defined as follows in a very sketchy way just to highlight the key points. This example is adapted from Fraenkel and Wallen (1993).

- Lexical: motivation is "the state or condition of having a strong reason to act or accomplish something."[51] It is related to the word *motive,* which originated from the Latin word *movere*, meaning *to move*[52].
- Conceptual: Students who are motivated to learn mathematics have the drive and enthusiasm to get started with the assigned work. They are eager to ask questions in class. Those who are not motivated take more time to copy down things or to start working on the given tasks. They do not seem to be aroused intellectually and emotionally during mathematics lessons.

---

[51] http://dictionary.reference.com/browse/motivation?s=t

[52] http://oxforddictionaries.com/

- Operational: Students' level of motivation to learn mathematics is measured by asking them to rate themselves on a 5-point scale on items such as: *I am excited about mathematics*; *I pay attention in mathematics lessons; I drive myself to complete difficult problems.*

It is hoped that explaining important education constructs using these three types of definition can place research about the affective domain onto much stronger theoretical foundations. Rigorous research can improve practice, so that the intended benefits of affective learning can be accomplished.

3 Importance of Affective Domain

There are three important curriculum reasons for focussing on the affective domain.

- Affect is part of many mathematics curricula worldwide because of the values attributed to strong student interest in mathematics. In Singapore, *Attitudes* is included as one of the five key components that help students become better problem solvers. In fact, the first hurdle in problem solving is that students must be motivated enough in the problems before they take the first step of trying to understand the problem. Without this motivation, they may give up before getting started.
- These affective goals contribute to students' holistic development, providing a balance against over-emphasis on academic results.
- Research has shown that some affective variables such as curiosity and grit can predict student achievement (e.g., Tough, 2013). This provides justification to strengthen students' positive beliefs and attitudes towards their learning.

It is also important to assess affective outcomes. Three reasons are offered below.

- It informs teachers to what extent the affective goals have been achieved. In Singapore, students are also awarded for affective outcomes. One of these awards is the Edusave Character Award

(ECHA), given to students who can demonstrate exemplary character and personal qualities through their behaviours and actions.

- By knowing the students' levels of interest, motivation, and other affective variables, teachers can plan follow-up activities for those who need help and encouragement in these areas.
- A truism in education is that "we measure what we treasure." When teachers measure these outcomes, they send a strong message to their students that these outcomes are important to possess. With constant reminder and assessment, more students may develop the positive affects. For example, if teachers regularly assess student effort, this experience will encourage more of them to put in greater effort in doing their homework and learn to attribute the quality of their output in terms of effort rather than innate ability (e.g., Dweck, 2006).

4 Attitudes: Meanings and Measurements

The Singapore mathematics curriculum uses the term *Attitudes* to cover five important affective constructs (Ministry of Education, 2012):

> Attitudes refer to the affective aspects of mathematics learning such as:
>
> - beliefs about mathematics and its usefulness;
> - interest and enjoyment in learning mathematics;
> - appreciation of the beauty and power of mathematics;
> - confidence in using mathematics; and
> - perseverance in solving a problem. (p. 19)

Studies about attitudes towards mathematics and its learning are very popular, especially among research students in mathematics education. Many instruments have been designed and validated to measure attitudes towards different objects. These instruments rely on self-reporting by the students. The popular ones are briefly mentioned below; see Tay, Quek, and Toh (2011) and textbooks on education measurement for further detail.

- Likert scale, administered in print or online (an emerging trend); this is by far the most popular method used; see Section 4.1 below.
- Semantic differential, based on the work of Osgood, Suci, and Tannenbaum (1957); see Section 4.2 below.
- Journal writing, using prompts about the attitude object. This is becoming popular in Singapore because it aligns with the communication goal of the curriculum; see Fan et al. (2008).
- Draw pictures about the attitude object, e.g., mathematicians, mathematics lessons, mathematics teachers; see Picker and Berry (2000), Wong (1996), and Wong et al. (2009).
- In-class (or in-situ) reflection about cognition, feelings, and classroom events at specific pauses during a lesson (Wong, 2000; Wong & Low-Ee, 2006; Wong & Quek, 2009); this is similar to the Experience Sampling Method of Csikszentmihalyi (1997). The basic scheme is shown in Figure 7.1. At each pause, a short checklist about the target variables (e.g., how do I feel?) is administered and collected in print or online. A noteworthy finding was that students and their teacher had different perceptions about the aim and feelings of the same lesson segment prior to each in-class reflection. Being aware of these differences, the teacher can plan subsequent lessons to address important gaps in cognition and affect.

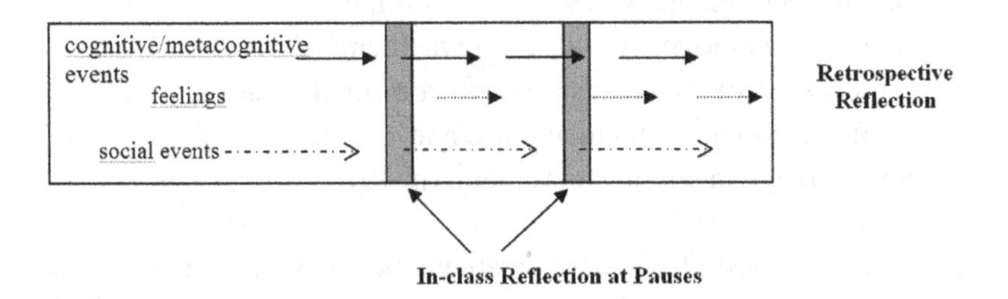

Figure 7.1. In-situ reflections during a lesson

- Combine mathematics item with affective response. For example, at the end of a mathematics problem, students are asked to rate how confident they are with their answer, whether they enjoy

solving the problem, or to what extent tackling the problem might help them solve similar problems in the future.
- Interviews using open-ended questions to collect qualitative data.

4.1 *Likert scale*

A Likert item is a statement which students are asked to indicate their degree of agreement on a number of points. The 5-point scale is widely used: *Strongly disagree, Disagree, Neutral, Agree, and Strongly agree*. Several items are usually written to describe the same construct, and they are classified into categories called *scales* or *dimensions*. The internal consistency of the items within a scale is assessed using Cronbach's alpha, and the validity of the scales verified using complex statistical analyses, such as factor analysis, Item Response Theory (IRT), and Structural Equation Modelling (SEM), which will not be discussed here.

Likert scales that measure attitudes towards education objects usually cover the following four dimensions:

- Ease, e.g., Mathematics is easy.
- Utility, e.g., Mathematics is useful.
- Confidence (self-esteem) about learning and doing mathematics, e.g., I am good in mathematics. It is important not to confuse self-rating of confidence with actual performance.
- Emotions, e.g., I am nervous when doing mathematics problems.

There are numerous Likert scales for measuring student attitudes towards mathematics, and the two widely cited ones are the *Fennema-Sherman Mathematics Attitudes Scales* and Sandman's *Mathematics Attitude Inventory*. TIMSS 2011 measured attitudes using a 5-item scale on *Liking Mathematics* and a 7-item scale on *Confidence in Mathematics* for both Grades 4 and 8, and a 6-item scale on *Students Value Mathematics* for Grade 8 only (Mullis et al., 2012).

The SMAPP project has produced a new instrument called the *Attitudes toward Learning Mathematics Questionnaire*. It consists of six scales with four items per scale; see Table 7.1.

Table 7.1

SMAPP: Attitudes toward Learning Mathematics Questionnaire

Scales	Items
Usefulness	1. Mathematics is important. 2. I think mathematics is useful in solving real world problems. 3.* I think mathematics is useful only for tests. 4. Mathematics helps me to understand reports and advertisements about prices, sale, percentages etc.
Confidence	1. I am good at using mathematics to solve real-life problems. 2. I am confident in solving mathematics problems. 3. I find mathematics easy. 4.* I am not good at giving reasons in mathematics.
Enjoyment	1. I enjoy doing mathematics. 2.* I find mathematics boring. 3. Overall, I have good feelings about mathematics. 4. Solving mathematics problems is fun to me.
Check solutions	1. When I know I have made a mistake in solving a problem, I will try to find out why. 2. After I have solved a problem, I will go through the solution again and check if I have made any mistakes. 3.* Once I have worked out an answer to a problem, I do not check my answer. 4. After I have solved a problem, I will ask myself if the answer makes sense to the given problem.
Multiple solutions	1.* I do not like to think of other ways to solve the same problem. 2. I often figure out different ways to solve mathematics problems. 3. I try to understand the different solutions given by my classmates. 4. After I have solved a problem, I will look for other methods to solve it.
Use of IT	1.* I do not like to use the computer to learn mathematics. 2. I can learn mathematics from playing computer games. 3. IT (Information Technology) has been helpful to my mathematics learning. 4. Mathematics software (e.g., graphing) helps me to learn mathematics.

The first three scales (Usefulness, Confidence, Enjoyment) encompass three of the four common dimensions of attitudes mentioned above; two scales deal with beliefs about problem solving; the last one is about ICT use, which was a major component of the study; see Chapter 6, Section 8. It was validated based on responses from about 750 Grade 7

Singapore students using a 9-point scale (Wong & Chen, 2012). For school use, the standard 5-point scale is adequate, but further validation may be necessary. The negative items (with asterisks) are to be reverse-scored before the summated score for each scale is computed.

4.2 *Semantic differential*

This method is not as popular as the Likert scale, but it is very easy to use, especially for students who may have difficulty understanding Likert statements due to their weak literacy. A set of about 10 bipolar adjectives about the attitude object is produced, and the polarities of the adjectives are reversed in a random order to prevent response set. Students indicate their *immediate* response to each bipolar adjective by circling a number or a mark placed between the two ends of the bipolar pair. Typically, seven points are used. An overall score is obtained by adding the numbers that correspond to the positive polarities of the adjectives. In general, strong correlations exist between attitudes measured using semantic differential and Likert scale. An example is shown below.

MATHEMATICS

1.	easy	— — — — — — —	difficult
2.	boring	— — — — — — —	interesting
3.	beautiful	— — — — — — —	ugly
4.	creative	— — — — — — —	mechanical
5.	very important	— — — — — — —	not important
6.	meaningless	— — — — — — —	meaningful
7.	for all	— — — — — — —	for the talented
8.	not enjoyable	— — — — — — —	enjoyable
9.	complicated	— — — — — — —	straightforward
10.	useful	— — — — — — —	useless

5 Research Issues about Correlations

Research about affective variables has a long history. A major review was undertaken by Aiken (1970), and since then, many narrative reviews and meta-analyses have been published. A vast majority of these studies involve correlation coefficients. This section examines six issues about correlational studies. Readers of these studies should be aware of these issues to avoid making wrong inferences from the findings.

First, a moderate positive correlation normally exists between mathematics achievement and attitudes towards mathematics or its learning, when the data are analysed at the student level. This correlation is around 0.3, suggesting a shared variance of 10% between achievement and attitudes.

Second, there are at least seven ways to interpret correlations. In the following discussion, let *A* be the measure attitudes, and *P* the achievement.

a) *A* causes *P*, $A \rightarrow P$. Enjoyment of mathematics causes high achievement; e.g., students who like mathematics tend to have higher achievement.
b) *P* causes *A*, $P \rightarrow A$. High achievement increases enjoyment of mathematics; e.g., those who do well in mathematics say they like mathematics.
c) *A* causes *P*, and *P* causes *A*, $A \leftrightarrow P$ (bi-directional causation). This leads to a cycle of causation. In this case, think of a *consistent* rather than *causal* relationship; e.g., students who like mathematics tend to have high achievement, and those who do well tend to like the subject.
d) *C* is the cause of *A* and *P* and the correlation between *A* and *P* is said to be *spurious*; see Figure 7.2(a). For example, enjoyment of mathematics and high achievement may be caused by effective teaching.
e) *D* is a *mediating* factor of the influence of *A* on *P;* see Figure 7.2(b). In this case, there is no *direct effect* from *A* to *P*. For example, enjoyment of mathematics affects the amount of effort put into learning and this effort gives rise to higher achievement.

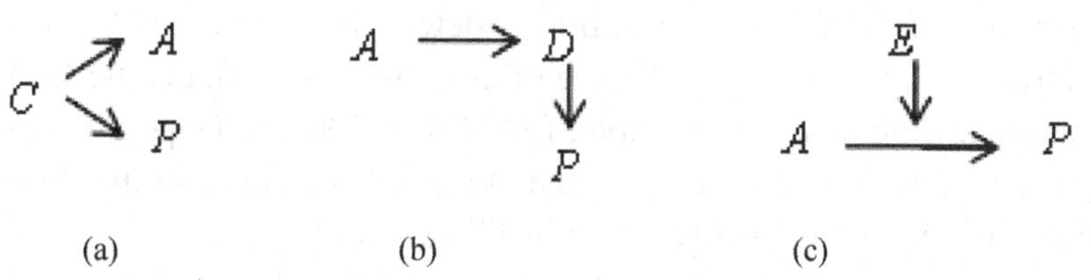

Figure 7.2. (a) *Spurious* correlation (b) *Mediating* factor D (c) *Moderating* factor, E

f) E is a *moderating* factor of the correlation between A and P; the correlation is changed when E is considered. See Figure 7.2(c). Another way to show moderating factor is $E \rightarrow (A \rightarrow P)$. For example, the correlation between A and P may be different for boys or for girls; in this case, gender is said to be a moderating factor.

g) When multiple causes are involved, each cause may contribute differently to the effect. The causes may have different pathways to the effect. It may be *additive* or *interactive:*
   i) In the additive way, $P = A1 + A2$. Each cause is, in itself, sufficient to produce the effect so that the effects are cumulative. For example, high achievement may be due to positive self-concept and enjoyment of the subject. Note that more than two causes may be involved.
   ii) In the interactive way, $P = A1 \times A2$. Each cause is necessary but not sufficient, in itself, to produce the effect. For example, positive self-concept interacts with motivation to produce high achievement. Note that more than two causes may be involved.

It is widely known that *correlations do not always imply causation*, and only cases (a) and (b) above suggest simple causation. Astute readers should be able to notice this violation in claims made about education findings. To make causation claims, the researcher should propose and justify plausible mechanism that might explain the correlation (Morrison, 2009).

Third, student attitudes towards mathematics are found to deteriorate over time. This is usually inferred from cross-sectional data. For example, in TIMSS 2011, the percentages of students liking mathematics were 50% and 25% for Grades 4 and 8 students respectively, leading the authors to conclude that "these attitudes deteriorate over time" (Mullis et al., 2012, p. 325). However, the use of cross-sectional data could lead to inconsistent findings. For example, for TIMSS 2007, Singapore Grade 4 pupils had more positive attitudes than Grade 8 students (Mullis, Martin & Foy, 2008), but this was reversed in TIMSS 2011.

To provide more convincing evidence of the decline in attitudes, researchers should conduct longitudinal studies. The SMAPP project was a recent attempt to do so. About 850 Grade 7 Singapore students took the Attitudes toward Learning Mathematics Questionnaire in March and then in September 2011. They had neutral attitudes in March (mean score 6.15 in a 9-point scale) and more negative attitudes in September (mean score 5.66), although the drop of 0.5 point was not statistically significant. Boys were found to have statistically more positive attitudes than girls at each administration. This gender difference is consistent with findings from other studies; for example, the US National Research Council and the Institute of Medicine (2004) noted that "[C]ompared to males, on average females typically rate their math competencies lower … and consider math less relevant to their future" (p. 76). Hence, teachers should find ways to arrest this decline in attitudes. Some strategies are recommended in Section 7 below.

Fourth, correlations depend on the unit of analysis. In general, correlations at individual level are smaller than those at group level, when the same data are aggregated by groups such as schools, regions, or countries. The sign of the correlation may also change. This is an example of the *ecological fallacy*. Three examples of this fallacy are given below to emphasise its prevalence in education.

- Using TIMSS data, Wilkins (2004) reported that correlation between mathematics self-concept and scores was positive and statistically significant (r = .11) at the student level but negative and statistically significant (r = -.58) at the country level.
- Using PISA 2003 data, Lee J. (2009) reported that within-country correlations of the same two variables (mathematics self-concept and scores) ranged from -.05 to .55, but between-country correlation was -.45. This value is of similar magnitude and sign to Wilkins', suggesting that certain mechanism at country level might explain both findings.

- Loveless (2006) found that at the national level, student *happiness* was inversely related to mathematics achievement!

Given that conflicting conclusions may be made based on correlations, it is not clear which correlations (individual or aggregated) should be used by policy-makers and teachers as evidence to support changes in policy and practice. Perhaps, some of the factors in the situated socio-cultural framework (Figure 1.3) may shed some lights on correlations at the national level.

Conversely, correlation based on a particular aggregated level, say, schools, should not be used to make inferences of the same relationship at the individual level (lower level of aggregation) or the national level (higher level of aggregation). One would need to justify any inferences made across levels with plausible underlying processes; unfortunately, this is seldom found in published work.

Fifth, as noted earlier, many studies use student self-reporting. Students are known to give socially desirable responses which are affected by cultural contexts. This affects the validity of these studies.

Finally, it is important to help students become metacognitive (aware of) their attitudes and beliefs and to guide them to change one belief at a time in the positive direction, thus integrating attitudes and actions. This is a long-term intervention, and the current practice of measuring student attitudes at only the pre- and post-stage in intervention studies does not give detailed information about any progress made. It is necessary to design easy-to-use attitude checklists that can be administered regularly. Online administration, as trialled in the SMAPP project, can simplify this process, and this will provide both students and teachers with timely data as assessment for learning.

6 Motivation: To Teach is To Sell

According to Colman (2006), motivation is "a driving force or forces responsible for the initiation, persistence, direction, and vigour of goal-directed behaviour" (p. 479). Two elements are highlighted in this definition: there must be goals to satisfy certain needs and effort taken by students to achieve the goals. Theories on motivation differ in the focus they place on the types of goals and strategies used to attain the goals. Six broad categories of theories are noted below:

- Cognitive theories focus on finding meanings in the tasks because students have a strong need to make sense of what they are learning and its relevance to their daily life. This is an important intrinsic motivator.
- Affective theories focus on emotional needs, such as having fun, being inspired, and feeling happy and confident. The symptoms for poor motivation are boredom, anxiety, indifference, and avoidance.
- Behaviourist theories deal with examinations and grades, tangible rewards, punishment, rule compliance, and reinforcements. The focus is on measurable outcomes of learning. These extrinsic goals can be achieved by applying behaviourist laws and they seem to work on a short-time basis and for routine tasks (cf. Pink, 2009).
- Socio-cultural theories emphasise the needs to belong to social and cultural groups, which can be satisfied when there is strong rapport between teachers and students as well as among students.
- Humanistic theories concentrate on the holistic development of the students, covering all the needs mentioned in Maslow's theory of human motivation: physiological, safety, love or belonging, esteem, and self-actualisation. Even spiritual needs have been mentioned for mathematics education (Winter, 2001).
- Neuro-science theories explain changes in the brain as a result of learning experiences. Students who understand that their brain can be wired differently as a result of learning have shown to become more motivated (Dweck, 2006).

It is obvious from the above brief survey that motivation is a multi-dimensional construct. Thus, it is important to consider the many causes of lack of motivation and engagement in the study of mathematics, especially beyond lower primary levels. Pink (2013) considered this issue from the sale perspective: teachers are trying to *sell* their lessons so that the students who invest energy, time, attention, and effort "will be better off when the term ends than they were when it began" (p. 39). To succeed in this sale, the teacher must deliver quality lessons (*products* and *delivery*), establish good rapport (*customer relationship*), and continue to support the clients (*feedback and follow-up*) to ensure satisfaction (*improved learning*). A multi-dimensional approach is called for, and the following *M_Crest* framework is one tool that can facilitate this.

7 The *M_Crest* Framework

In 2013, I introduced this framework to apply motivation theories to plan mathematics lessons that can engage the students. It is evident that no single motivator can work for all students all the time and for every topic. Hence, it is necessary to integrate these motivators to suit the ever changing teaching situations. The essential features of this framework are explained below, with reference to earlier chapters. A longer paper with slightly different coverage can be found in Wong (2014). The acronym, *M_Crest,* is created to remind teachers that they can help all students to rise to their potential *crest* in mathematics.

The six motivators covered in the framework are: *Meaningfulness, Confidence, Relevance, Enjoyment, Social relationships,* and *Targets.* Four out of these six motivators are also mentioned in the Singapore mathematics curriculum: *meaningful, confidence, relevant,* and *fun.*

7.1 *M = Meaningfulness*

Students have a strong need to make sense of the mathematics they are learning. Hence, helping them to do so is a very important motivator. Unfortunately, this is unlikely to happen if the lessons focus primarily on practising routine skills, without covering the underlying conceptual understanding and mathematical reasoning. This leads many students to believe that mathematics is meaningless and tricky. This malaise also affects intelligent people during their younger days. For examples, Bertrand Russell (1872 – 1970) lamented that his tutor became angry when he could not recite the expansion of perfect square in words, and Carl Jung (1875 – 1961) was terrorised by incomprehensible algebra but was too afraid to ask questions. By realising that these negative experiences may affect their students, the teacher should strive to make mathematics meaningful.

The promotion of meaningful learning in mathematics has a long tradition, at least in the US. Half a century ago, Brownell (1947/2004) defined *meaningful arithmetic* as "instruction which is deliberately planned to teach arithmetical meanings and to make arithmetic sensible to children through its mathematical relationships" (p. 10). As discussed in Chapter 2, when teachers depend solely on words and symbols to communicate mathematics, many students fail to understand relationally what is taught. When students fail to grasp the meanings conveyed by mathematical symbols, they generalise incorrectly. The two examples below are only the tip of the iceberg of algebraic errors:

- ab means a times b, whereas 23 means 20 + 3 and not 2 times 3.
- $2(x + y) = 2x + 2y$, but $(x + y)^2$ is not equal to $x^2 + y^2$.

The strategies discussed in Chapters 2 to 4 can be used to activate this motivator: multiple representations, patterns, examples and non-examples, and Dienes' variability principles.

7.2 *C = Confidence*

Students need to gain confidence in their abilities to learn complex mathematics and to solve mathematics problems. When their confidence rises, they are more willing to tackle challenging learning tasks.

A general belief is that this need can be achieved when students experience success in their work. Bruner (1960) noted that this self-confidence should "come from knowledge of a subject" (p. 65). Thus, teachers must ensure that their students really master the requisite knowledge rather just feel good about completing easy tasks.

The students themselves need to realise that they should not expect to succeed immediately in every learning task because deep learning comes from meeting challenges rather than staying in error-free comfort zones. They also need to understand that one can learn from mistakes. Most of the mistakes made in initial learning are temporary because they can be corrected. They are also specific to the types of problems rather than reflective of innate mathematical inability. Of course, teachers should provide timely, specific, and encouraging feedback so that students can rectify their mistakes and move forward. Getting them to work in the ICT tutee mode as explained in Chapter 6 is a more engaging way to inculcate this mindset.

Some teachers believe that praising students will raise their confidence level. However, giving vague praises (e.g., "good work" or "good job") actually creates *praise junkies,* who then depend on external praises rather than develop their own self-confidence and sense of achievement (Kohn, 2001). Similarly, praising ability (e.g., you are really smart) can be detrimental because the students would avoid difficult work in order not to undermine their sense of being smart. Instead, Dweck (2006) found that praising students for effort (e.g., you have worked hard on this) inculcates a growth mindset such that the students are more willing to try harder problems and to be more engaged in their learning. This is an effective routine to bolster students' confidence.

Conversely, students who experience more failures than successes are likely to be demotivated by avoiding taking risks to complete the tasks or giving up completely. This becomes a vicious cycle of failure (Butterworth, 1999): bad performance → external discouragement → internal discouragement → anxiety → avoidance → no improvement. An effective strategy to motivate low performing students is to give them graded problems so that successes with simple problems may stimulate them to tackle the more complex ones. They can keep a record of their little successes as a powerful indicator of progress. Also teach them how to use positive self-affirming talk, such as "I will be able to solve it" or self-interrogative questions such as "can I solve it?"

Teachers should avoid giving negative comments because such external discouragement may be internalised leading to anxiety and avoidance. Indeed, teachers need to build a supportive environment in which mistakes, uncertainties, and risks are seen as springboard to deepen learning.

7.3 *R = Relevance*

Two kinds of relevance are pertinent here: cognitive and emotional. Cognitive relevance refers to seeing values in learning mathematics because it can be applied to everyday situations, for the learning of other subjects, and in future careers. Many students express this need for cognitive relevance when they ask questions about why they are studying certain topics. Answers to such query are found in mathematics applications and modelling, which are discussed in Chapter 5.

TIMSS 2011 (Mullis et al, 2012) found that nearly half the international sample of Grade 8 students believed that mathematics had values, and they had the highest average achievement. The 15% who did not value mathematics had the lowest average achievement. Although this does not imply that getting students to appreciate the values of mathematics will raise their mathematics achievement, the relationship is strong enough to justify utilising cognitive relevance as an important motivator.

Cognitive relevance is also about addressing dissatisfaction when one does not understand something important. When students feel that they have missed important points in lessons or have made mistakes, they will be motivated to fill these gaps by seeking further explanations from the teacher or friends. Teacher can exploit similar cognitive relevance by mentioning to the class how what follows will build on previous lessons.

Emotional relevance focusses on ensuring that learning experiences satisfy the emotional needs of students. This is very important because "[e]motions are the gatekeepers to the intellect" (Pete & Fogarty, 2003, p. 13). Emotionally relevant tasks satisfy students' curiosity, are fun to do, and lead to strong sense of success. This leads to the next motivator.

7.4 *E = Enjoyment*

Enjoyment in doing a learning task is a powerful factor to move students towards its completion. This is especially effective if the enjoyment also satisfies their curiosity about the environment or mathematics. This is an important type of intrinsic motivation related to whether students like or dislike learning mathematics. According to Butterworth (1999), enjoyment plays the critical role in the virtuous circle of success: good performance → external encouragement → internal encouragement → enjoyment of mathematics → more mathematics → better understanding.

Enjoyable lessons can be delivered using the strategies below.

First, teachers can include in their lessons a wide variety of surprises, tricks, puzzles, riddles, magic, games, competitions, humour, cartoons, and other activities to arouse student interest and curiosity in the topic. Introducing a new topic with some of these activities can make the learning more memorable and engaging. Several examples are given below; brief comments (in italics) are added to highlight the relevant mathematical ideas.

- Use six sticks of the same length to form four equilateral triangles. *Introduce 3-D object and check assumption.*
- Calculate: 1 × 1, 11 × 11, 111 × 111, and so on. When does the pattern break down? *Pattern and place value, for example, think of 111 × 111 as 111 × (100 + 10 + 1). It breaks down after 111,111,111.*
- Three girls had lunch at a restaurant. The total bill was $30, so each girl paid $10. The cashier realised that the bill should be only $25, so she asked the waiter to return $5 to the girls. Each girl took back $1, and they gave the waiter the remaining $2 as a tip. Now, each girl paid only $9, giving a total of $27. The waiter had $2, so where was the missing dollar? *Be systematic.*
- A number like 133 is said to be "ordered" because each digit is not smaller than the digit to its left. How many 3-digit numbers are ordered? *Understand the role of definition and be systematic.*
- By taking only one measurement, determine the area of an annulus. *Pythagoras' Theorem.*
- In a country, all parents want to have a girl. Every couple keeps having children until they have a daughter; then they stop. What is the proportion of girls to boys in this country? Assume that it is equally likely to have a girl or a boy and there are no multiple births. *Probability is not required;* see Poundstone (2012).
- A spaceship reported that the temperature in outer space was -40°. Was it in Fahrenheit or in Centigrade? *The special case where both temperature scales give the same value.*

Second, the teacher should deliver the topic or problem with enthusiasm so as to transmit the same feelings to the students. This hypothesis was supported by a large-scale study involving about 1500 Grades 7 and 8 students in Germany (Frenzel et al., 2009). They found that "a teacher's enjoyment during teaching has positive effects on student enjoyment via observable enthusiastic teaching behaviour" (p. 712). In contrast, when a topic is taught in a business-as-usual, mechanical manner devoid of excitement, the students quickly lose interest. This perception is also supported by a survey of about 270 pre-service and practising teachers (Lim & Wong, 1989). These teachers rated very highly the ability to convey an enthusiasm for mathematics to the students (mean 3.8 out of 4). They also believed that teachers should have genuine interest in mathematics.

Third, learn about what the target students find enjoyable because enjoyment is subjective and varies with past experiences, current needs, and future expectations. To gather this kind of information, teachers need to have strong rapport with their students, and this is the next motivator to be discussed.

7.5 *S = Social Relationships*

Classroom teaching takes place between teacher and students. It is a social activity. A strong motivator to entice students to learn under this interactive context is to ensure that the social relationships in the classroom are cordial and supportive. Some ways in which this motivator can best function are:

- Inculcate a sense of belonging to the same class so that the students will support one another in trying to master the mathematics. This includes helping students develop social skills required for productive group work.
- Listen carefully to students' solutions and answers, which may be wrong, as windows into their thinking rather than rejecting them outright or putting red crosses all over their homework. This sends a positive message that students are respected as individuals. Indeed, there are many stories where students turn over a new leaf

because their teachers care about their social and emotional welfare beyond merely teaching the subject.

- In a bilateral study, Singapore primary school students ranked being patient as the most important personal quality of their *best* mathematics teachers (Wong et al., 2009). This was followed by qualities such as being humorous, helpful, and dedicated.
- An emerging trend is to use social media to establish these relationships because this approach is much appreciated by both students and teachers who are digitally savvy. However, this should build on rather than replace face-to-face interactions, since subtle, impactful messages can only be conveyed through body language.

7.6 *T* = *Targets*

Setting clear and achievable targets is a powerful motivator. These targets may be immediate, short-term, or long-term.

Immediate targets refer to what have to be completed within a lesson or for homework. Doing so is motivating only if the students understand the learning objectives to be achieved. These objectives should be stated in student-friendly language and shared with them. One effective approach is to write down the lesson objectives on the board and tick them off whenever they have been covered. This allows students and teacher to keep track of the teaching and learning process.

Short-term targets take longer time, perhaps weeks and months, to accomplish. These are often included in the scheme of work or unit plan spread over several weeks, but they are seldom conveyed to the students. There is merit in giving students a broad plan so that they are aware of the expectations to be achieved at the end of a significant period of time and not be unduly distressed if they fail to master something in the short term. This conveys the message that deep learning takes time, and minor failures or hindrances along the way should not be too deliberating.

Long-term targets refer to broad goals for the whole mathematics curriculum. Unfortunately, many teachers lose track of these goals when they focus primarily on *completing the syllabus* for individual lessons. At the school level, teachers should meet to brainstorm how their lessons can contribute to these goals from one grade level to the next. Becoming aware of this might prompt some of them to remind their students about these important broad goals, and hence motivating them to consider long-term aspirations.

7.7 *Combining the Motivators*

The six motivators in the *M_Crest* framework cover different types of motivation to learn. They can be applied separately to fairly focussed situations. They become truly powerful only when applied in appropriate combination. One such combination is as follows:

- *E:* begin with an enjoyable activity;
- *M*: stress conceptual meaning;
- *T:* practise skills to achieve well-defined immediate goal;
- *C:* gain confidence through mastery of skills;
- *R:* show cognitive relevance of the mathematics contents and reasoning; and
- *S:* these take place in a socially positive environment.

Teachers are urged to experiment with different combinations of these motivators and find out for themselves which ones work well for their students. Sometimes students may begin working on learning tasks without much enthusiasm, but incorporating some of these motivators along the way may spark their interest, resulting in positive changes in attitudes towards learning.

8 Concluding Remarks

This chapter explores many aspects of the affective domain: its role as desirable learning outcome, its link to the cognitive domain, its measurement, and related research. Because motivation to learn mathematics is a critical issue in many countries, the *M_Crest* framework is offered here as a powerful and practicable way to energise students' intentions to study mathematics. However, knowledge of such a framework and an espoused willingness to use it is, as noted by Goethe, not enough; teachers must act on it to help students acquire these affective resources. Finally, the students must develop the metacognitive competence to strategically use their affective and cognitive resources to become better problem solvers and mathematics learners. The next chapter will deal with these functions of metacognition.

Chapter 8

Metacognition: Strategic Use of Cognitive Resources

Success in learning and problem solving requires that students can strategically apply their cognitive resources to the learning tasks. This includes retrieving the appropriate knowledge, monitoring its use and effects, changing course if necessary, and reflecting on the outcomes against the objectives. These processes collectively constitute metacognition. The Singapore mathematics curriculum refers to two aspects of metacognition: self-regulation of learning and monitoring of one's own thinking. This chapter proposes two frameworks for metacognition, one for each of these two aspects, and elaborates on several strategies which can facilitate the development of these two metacognitive processes in order to improve mathematics learning and problem solving.

When the mind is thinking it is talking to itself. Plato[53]

Learning without thinking is labour lost. Confucius (*Analects*, Chap. 2, Verse 15, 学而不思则罔)

1 Metacognition: Meanings and Importance

According to *Dictionary.com*[54], when the prefix *meta-* is added to a discipline, such as mathematics, the result is a new subject that

---

[53] http://www.brainyquote.com/quotes/quotes/p/plato159586.html

[54] http://dictionary.reference.com/browse/meta?s=t

investigates the original one in a more abstract or deeper way. It gives the sense of *going beyond* something. For example, meta-mathematics is a study of the foundations of mathematics, going beyond specific mathematics. This notion of going beyond or deeper is also applied to actions: meta-analysis is "analysis of analysis" (Glass, 1976), meta-learning is "learning about learning," and meta-cognition is "thinking about thinking." When an action is meta-ised, new concepts and methodology are applied to the original action. For example, under meta-analysis, the studies to be analysed further are examined to yield effect sizes, and these are then compared statistically, so that a summary can be made about the original studies.

In the case of metacognition, Flavell, who coined this term in a seminal paper published in 1976, noted how metacognition might differ from cognition:

> one's knowledge concerning one's own cognitive processes and products ... active monitoring and consequent regulation and orchestration of these processes in relation to the cognitive objects or data ... in the service of some concrete goal or objective. (p. 232)

While the difference is not unequivocal and absolute, it is still useful to keep it in mind because the ways of teaching metacognition are not always the same as those for cognition. This will be covered in Section 3.

In mathematics problem solving, Schoenfeld (1987) identified three aspects of metacognition: self-awareness, control, and belief about one's cognition. Fogarty (1994) used the term *metacognitive reflection* to cover planning, monitoring, and evaluating of learning carried out by the students. Other researchers distinguish between metacognitive knowledge, metacognitive skills, and metacognitive beliefs. The US National Research Council (2000) included metacognition as one of three overarching learning principles and gave the following definition:

> A "metacognitive" approach to instruction can help students learn to take control of their own learning by defining learning goals and monitoring their progress in achieving them. (p. 18)

The Singapore mathematics curriculum framework refers to two aspects of *metacognition;* one of it is "monitoring of one's own thinking" and the other is "self-regulation of learning." The recurring ideas in the study of metacognition include awareness, control, monitoring, self-regulation, and reflection.

The 2013 Yearbook of this series consists of 15 chapters contributing to research and praxis of metacognition and reflection in mathematics education (Kaur, 2013). Other reviews about metacognition can be found in Desoete and Veenman (2006) and Waters and Schneider (2010). This expanding literature underscores the importance of metacognition as an essential component of active learning and effective problem solving. The following sections reiterate some key ideas in my earlier papers on metacognition (Wong, 2002a, 2013b) and introduce new materials based on my new learning and consultation with Singapore schools.

2 Self-Regulation of Learning

This aspect of metacognition derives from the *learning to learn, learning styles, learning strategies,* or *study behaviours* reform. The reform was popular in the 1980s among language educators and cognitive psychologists, and as a result many generic programmes to teach students how to learn are developed, especially at post-secondary level (e.g., Biggs, 1987; Crawford et al., 1998). Unfortunately, these efforts have not penetrated into mathematics education, because mathematics teachers and researchers are more concerned about mathematics pedagogy than generic strategies. This is evident from the fact that most of the terms mentioned above do not appear in the indexes of recent reviews of mathematics education. Mathematics teacher education also hardly includes these strategies in their methodology courses.

Good mathematics learners, however, often display effective study behaviours that they have developed on their own. These behaviours include setting short-term and long-term learning goals, being organised in study schedule, keeping accurate notes of mathematics and corrections of one's own mistakes, completing homework on time, seeking help and support when necessary, revising the work regularly, and enhancing memory using a variety of strategies. They "can develop sufficient insights into their own learning to improve it" (Wiliam, 2011, p. 146). The mathematics-specific strategies mirror good teaching, and cover behaviours such as thinking of alternative methods, looking for examples and non-examples, posing own mathematics questions, visualising mathematical properties, and reading more about mathematics. Similar ideas are offered by other educators, for examples, Arem (2010), Birken (1986), Cooke (2003), Corkill (1996), Margenau and Sentlowitz (1977), Mason (1999), Novak and Gowin (1984), and Ooten (2010). The long-running *Project for Enhancing Effective Learning* (PEEL), launched in 1985, aims to foster effective, independent learning among students by training them to enhance metacognition (Baird & Northfield, 1992). Their long list includes some of the strategies mentioned above.

Poor learners, on the other hand, are either not aware of these strategies or do not practise them consistently. This was pointed out by Buoncristiani and Buoncristiani (2012): "students make poor choices about how they study because they are unaware of alternatives and their consequences" (p. 128). Thus, teachers need to find out what strategies their students are currently using and consider how to help them acquire more effective ones. Without a varied repertoire of effective strategies, students would have nothing much to regulate about their learning.

2.1 *A framework about self-regulated learning*

A learning strategy is a student's way of acquiring new knowledge, remembering it, deepening it, and using it to solve problems. Different strategies can be used during lessons and after lessons to achieve different instructional foci. To be able to master a new learning strategy, students must understand its nature (metacognitive knowledge), practise it over and over again (metacognitive skill), monitor how well they have used it (metacognitive monitoring), and reflect on its effect (metacognitive reflection). With repeated practice, they develop their own metacognitive belief about which strategy will be beneficial for future learning. Table 8.1 illustrates how these ideas work for three selected learning strategies. This framework is an aid to make these metacognitive elements visible to the students.

Table 8.1

Metacognitive knowledge (MK), *metacognitive skills* (MS), *metacognitive monitoring* (MM), *and metacognitive reflection* (MR) *for three learning strategies* (LS)

LS	MK	MS	MM	MR
Monitor learning objectives	Important to know what the lesson is about.	Record objectives in note cards.	Tick off each objective when it is covered.	Which objectives are not yet learned?
Deliberate practise	Practice makes perfect.	Study worked examples. Submit homework. Do more problems.	Note worked examples that help in doing homework.	Covered all worked examples? Which type to remember?
Overcome mistakes	Mistakes as springboard of learning.	Classify mistakes and corrections.	Avoid repeating these mistakes when solving new problems. Revise notes regularly.	Which corrections have been mastered? Which mistakes are still troublesome?

Notice that metacognitive skill does not refer to specific mathematical rules; it refers to the ability to take appropriate actions based on the metacognitive knowledge. In general, there will be more than one metacognitive action for each learning strategy, as illustrated by the three entries under MS for *deliberate practice.* Many students believe that they have to practise by completing homework, but they do not know how to practise effectively. The entries under metacognitive reflection (MR) are questions that teachers can discuss with their students so that they can internalise these questions to guide their use of these strategies.

Weak students are also found to have poor long-term memory of what they have been taught, and efforts must be devoted to enhance their memory. According to TIMSS (2011) (Mullis et al., 2012), mathematics teachers internationally reported spending some of their class time asking students to memorise rules, procedures, and facts. About 37% of Grade 4 teachers and 45% of Grade 8 teachers reported doing this every or almost every lesson, even though the percentages vary widely across the countries. For Singapore, about 20% of the teachers at each level reported the use of this instructional activity, but the study did not collect data about the memory strategies taught.

Information processing theories suggest that memory improves with strong associations across ideas. This can be achieved by various means: recite the rule aloud, listen to mathematics in songs[55] or videos[56], write it down from memory and check that this has been done correctly, make up mnemonics, form mental imagery using concept mapping, creating posters, write their own notes, and re-do worked examples from memory.

The next section deals with the vexing problem of why many mathematics students do not ask questions in front of the class when they do not understand parts of the lesson, even though they may seek help from the teacher after class.

2.2 *Student Question Cards (SQC)*

It is generally believed that students will benefit from asking questions because this encourages meaningful learning, resolves doubts, promotes critical thinking, and satisfies the urge for curiosity, especially among young children. Thus, effective teachers value and encourage student questioning (Gibboney, 1998; Rothstein & Santana, 2014). Effective students ask questions at the opportune moments when they do not understand something during the lessons. However, many students in Singapore and other countries are found to be fearful or reluctant to do so in front of the class at these crucial moments. It seems that they:

- do not know how to ask questions that specifically describe their real learning difficulty; instead they ask the teacher to repeat the whole explanation from the beginning; this does not allow the teacher to give answers that can address their actual difficulty;

---

[55] http://www.songsforteaching.com/mathsongs.htm

[56] http://www.neok12.com/

- do not wish to look stupid in front of their friends by asking what may appear to be simple questions.

The rest of this section describes how the *Enhancing Mathematics Performance of Mathematically Weak Pupils (EMP)* project made an attempt to help a sample of upper primary and lower secondary school students overcome these two issues (Wong & Quek, 2009, 2010; Wong, 2012).

In mathematics, questions cover meaning, method, reasoning, and application. Sample questions for these four types of questions shown in Figure 8.1. An option is given for students to ask their own questions.

Meaning M1: What do you mean by …. M2: What is the difference between …. and …. M3: Can you use a diagram to show …. M4: Is this related to something you taught us before? (Your own question)	**Method** Md1: Can you show us how to do this problem in another way? Md2: Can you explain/show us this step (….) again? Md3: What will you do next? Md4: How do you know what method to use? (Your own question)
Reasoning R1: Why do you do that ….? R2: What happens if you change …. to ….? R3: Why do you say this is harder than that one? R4: Does this … still works if we …? (Your own question)	**Application** A1: Why do we study this topic (….)? A2: How do we use (….) in our life? A3: Do we need this if we want to become …? A4: Have you (teacher) come across this in your daily life? (Your own question)

Figure 8.1. Student Question Cards (SQC)

These questions are printed onto cards known as *Student Question Cards* (SQC). Each student is given a set of SQC. This tool addresses the above two issues in this way. First, students learn to ask specific questions by referring to the sample questions on SQC. Second, they are less likely to feel stupid because the questions are chosen from the given list in SQC. Through practice with SQC, students will be able to overcome these two obstacles and become better at asking their own questions. SQC serves as a scaffold, which will be gradually removed to promote metacognitive control.

To use these cards, the teacher conducts the lesson as normal but stops after about 10 minutes of explanation and asks the students to raise their card to indicate which type of questions they wish to ask at this *question time*. The teacher scans the cards raised and calls on one or more students to ask their questions, which the teacher then answers on the spot. This process is repeated two or three times per lesson, depending on the lesson plan.

As a concrete example, suppose the teacher has just explained the difference of two squares, $x^2 - y^2 = (x + y)(x - y)$ and a worked example like $9 - 25x^2$. She pauses and invites the class to ask questions. Some possible questions might be:

- *Meaning*: Can you show this on a diagram?
- *Method*: Can you show us how to do this with another example?
- *Reasoning*: What happen if you reverse the x and y, like $y^2 - x^2$? Do you get a negative value?
- *Application*: How do we use this in our daily life?

During the implementation of the EMP project, the participating teachers were allowed to decide how to select the students to ask questions during question times. Some teachers asked every student to choose a question while others invited only those who had not understood; some focussed on one type of questions, say *method*, in a lesson and moved on to other types in later lessons; some chose quiet students to encourage them to participate in the lesson. Most of the teachers changed their ways of using SQC over the one month trial period. For group work, students can discuss their questions before the group decides on which question to ask.

The teachers generally had difficulty giving immediate, cogent answers to questions about drawing diagrams to illustrate the rule and everyday life applications. These two areas require special attention in their professional development.

There were mixed responses from the teachers and students on this technique. On the positive side, the question times broke the monotony of typical lessons, encouraged participation from the quiet students, and alerted teachers to learning difficulties evident at the moments. Most of them found the cards easy to use but expressed neutral opinion about whether this helped them understand the lessons better. Some students enjoyed the activity, while others did not. Most of the teachers commented on the extra planning time and curriculum time required. However, it is also important to note that this trial was probably too short for the teachers to develop competency in a new teaching strategy.

Despite the constraints mentioned above, it is important to give greater opportunity to more students to ask rich questions in class to ameliorate the dominance of "teacher asks and students answer" practice, the *IRF* discussed in Chapter 4, Section 7, which may impede active learning from the students' perspective. As Einstein commented, "The important thing is not to stop questioning."[57] The purposes differ for the students and teachers: for students, questioning is to clarify doubts, to satisfy curiosity, and to improve learning; for teachers, encouraging students to ask questions is to augment their instructional repertoire. When students feel confident and uninhibited to ask their own questions in class, this is the beginning of self-regulation of active learning.

2.3 *Studies about learning strategies*

This section describes five studies about the learning strategies or study behaviours reported by mathematics students, in order to identify areas where self-regulated learning can be promoted and further investigated.

As part of my doctorate study (Wong, 1984), 216 Grade 11 students from seven state schools in Brisbane, Australia, responded to a 40-item questionnaire that covered the following aspects of study behaviours: study organisation, use of notes, problem solving, memorisation, reading, and preparation for tests. They rated how often they engaged in these behaviours when they studied mathematics on a 5-point scale: 1 = *Never*; 2 = *Seldom*; 3 = *Sometimes*; 4 = *Often*; 5 = *All or most of the time*.

The more frequently used behaviours were (means in parentheses):

- Pay attention to teacher's hints about which topics are likely to come up in a test and concentrate on them. (4.2)

[57] http://rescomp.stanford.edu/~cheshire/EinsteinQuotes.html

- Hand in homework on time. (4.1)
- Pay attention in class. (4.1)
- Pay attention to mistakes and their corrections so that they will not be repeated. (4.0)
- Find a favourable environment, e.g., a quiet room for studying maths. (4.0)
- Concentrate more on how to use a rule than on how to derive or prove it. (3.9)

The above behaviours are generic and effective students are expected to engage in them. In contrast, the less frequently used behaviours were:

- Try to make up different proofs to a result. (1.9)
- Depend on luck in tests rather than study for them. (1.9)
- Do maths puzzles or play maths games in spare time. (1.8)
- Read several different books on the same topic. (1.7)

These self-initiated activities were hardly used, which was not surprising given the typical mathematics learning culture in many schools. Finally, the active or constructivist behaviours were used sometimes or infrequently:

- Try to judge whether answers to exercises are correct or not before looking up the given answers. (3.3)
- Try to link different mathematical ideas together. (2.9)
- Compare notes with friends. (2.6)
- Try to think of different ways of solving a particular problem besides those given by the teacher or the textbook. (2.3)

The general picture that emerged from analysing mean scores was that these Australian students tended to adopt a traditional approach to study mathematics. However, students are likely to use these behaviours in different combinations called *profiles*; for example, a student's profile may include paying attention in class, judging correctness of answers, and doing puzzles. These individual profiles cannot be inferred from the mean scores alone. They can be determined using cluster analysis.

Cluster analysis is a set of statistical techniques used to place previously unclassified objects into groups based on a set of variables (e.g., Everitt, Landau, & Leese, 2001). It has been successfully used in education. A well-known example is the study by Bennett (1976). In that study, 468 primary school teachers were clustered into 12 types along an informal-formal-mixed continuum, and nearly 1000 of their students into eight types based on personality profiles. For mathematics achievement, students of all the types made gains in formal classes but under-achieved in informal or mixed classes. Other researchers have also applied cluster analyses to discover new insights about education, e.g., Borman et al. (2005), Beswick (2006), and Law (2007). This supports the use of this statistical technique to analyse study behaviour data.

The 40 items in the Brisbane study were first factor analysed into nine scales. Then the students were clustered into five groups based on these nine scales. The groups were characterised as follows (number of students in each group given in parentheses):

- *Conscientious* students (73); they employed most of the behaviours in their learning.
- *Independent* students (50); they were diligent but reported solving problems in different ways or reading more about mathematics than the other groups.
- *Cue-seeking* students (36); they were diligent but seldom checked whether the work made sense or not, scoring relatively high on paying attention to hints.
- *Minimal* students (29); they reported low frequencies on many behaviours; they expended little effort in their study.
- *Rote* students (28); they relied on memorisation and attention to notes and hints and were not particularly diligent.

On a common calculus test, the *Independent* group had the highest mean score while the *Rote* group the lowest. Thus, study behaviour profiles are associated with different mathematics achievement.

In 1988, a modified version of the above questionnaire was administered to 1318 Grades 7 to 12 students in Singapore (Wong, 1991). The results for the Australia and Singapore samples on selected items are compared in Table 8.2, where the items are arranged in descending order by the differences in mean scores (Singapore mean – Australian mean). Students from the two samples expressed similar use of most of these behaviours. The greatest contrast was about memorisation: the Singapore students rated this as their most frequently used strategy, whereas the Australian students seldom used it. This appears to match the stereotypes about how students in the East and West study. At the other end, the Australian students were more likely than the Singapore students to keep notes of mistakes and corrections, although the latter also used this strategy some of the times.

Table 8.2

Study behaviours: Singapore vs. Australian samples

Item	Singapore	Australia	S - A
Memorise formula	4.2	2.6	1.6
Compare notes and solutions with friends	3.1	2.6	0.5
Try different methods of solving same problem	2.6	2.3	0.3
Do maths puzzles in spare time	2	1.8	0.2
Complete homework on time	4.1	4.1	0
Pay attention in class	4.1	4.1	0
Try to judge whether answer is reasonable or not	3.2	3.3	-0.1
Borrow maths books from library	1.5	1.7	-0.2
Pay attention to teacher hints	3.8	4.2	-0.4
Concentrate on how to use rule rather than on why it works	3.2	3.9	-0.7
Keep notes of mistakes and corrections	3.2	4	-0.8

Wong and Veloo (1997) assessed beliefs about learning mathematics held by about 400 Grades 7 and 8 students in Brunei Darussalam after they were taught using the multi-modal teaching strategy over six months from March to September, 1996. The instrument, *Beliefs About Learning Mathematics* (BALM), consisted of pairs of statements, placed in the semantic differential format (see Chapter 7). Students indicated their belief on a 5-point scale between the positive and negative ends of each pair of statements. A typical pair of statements is shown below.

I do only those exercises set by my teacher.	— — — — —	I try additional exercises besides those set by my teacher.

This forced-choice Likert-type format is less prone to acquiescence response set as in the *agree vs. disagree* Likert format (Converse & Presser, 1986). These scales had moderate internal consistency with Cronbach's alphas from 0.59 to 0.68, suggesting this format is acceptable for research work. The results of the eight study behaviour items are given in Table 8.3, where the items are arranged according to the pre-test means. These Bruneian students generally sought help to understand the lessons, showed some preference for hands-on activities, but were not keen on doing extra work on their own. There were slight declines in these beliefs after the intervention. The largest change (only a difference of 0.3 in mean scores) was that the students expressed greater difficulty in remembering the rules in September compared to in March, probably because there were more rules to remember in the latter months.

Table 8.3

Beliefs about study behaviours reported by Brunei students

Negative End	Positive End	Pre	Post
When I do not understand some maths in class, I keep quiet and forget it.	When I do not understand some maths in class, I ask the teacher or other students.	4.22	4.23
When the teacher asks a question in class, I do not bother about it.	When the teacher asks a question in class, I try to figure out the answer.	4.14	3.96
In doing problems, I try to apply the rule I first think of.	In doing problems, I try to understand it before thinking of a rule.	4.11	3.95
My notes about maths are sketchy and untidy.	I keep good notes about maths.	3.56	3.36
I do not like to learn maths by doing activities; I prefer to follow the rules.	I like to learn maths by doing activities, e.g., cutting, pasting, paper folding, playing games.	3.49	3.56
I do not bother about the life stories of mathematicians.	I like to know the life stories of mathematicians.	3.39	3.29
I find it difficult to remember all the rules in maths.	I have no difficulty remembering the rules.	2.97	2.68
I do only those exercises set by my teacher.	I try additional exercises besides those set by my teacher.	2.86	2.86

The EMP study (see Section 2.2) also examined the study behaviours of 147 students (68 Grade 4 and 79 Grade 7). They answered a questionnaire comprising 30 activities similar to those used in the three studies above. The students rated each activity on:

- how frequently they had used each activity using a 4-point scale: 1 = *Never*; 2 = *Occasionally*; 3 = *Often*; 4 = *Always*;
- how helpful they found each activity using a different 4-point scale: 1 = *Waste of time*; 2 = *Not so helpful*; 3 = *Helpful*; 4 = *Very helpful*.

The results are reported in Table 8.4. In general, traditional activities were rated highly in occurrence and helpfulness: pay attention, hand in homework, ask for help, and memorise things, do a lot of exercises (only assigned ones). The more metacognitive behaviours (5 items indicated with asterisks in Table 8.4) were used only occasionally and considered as not helpful: write reflections in a journal, make own notes, keep track of corrections to mistakes, plan own schedule of studying mathematics, and explain things to others. ICT use was rare and rated as not so helpful.

The fifth and final study was conducted by Ee (2014) to probe the study behaviours of a group of 235 students who were studying an Engineering Mathematics module in a Singapore polytechnic. Her *Student Study Questionnaire* comprised 57 items, which were modified from the above studies and that of Biggs (1987) to take into account the polytechnic environment. These items were conceptualised to cover the following four constructs, and the students responded to each item about its frequency of use and helpfulness:

- Conscientious effort (15 items)
- Metacognition (16 items)
- Involving others (15 items)
- Resources (11 items)

Table 8.4

Frequency (F) and Helpfulness (H) of study behaviours from EMP Project

Items	F	H
Pay attention to teacher when he or she explains things.	3.05	3.39
Hand in homework on time.	2.97	2.97
Listen to others explaining things in class.	2.90	2.88
Do something nice after I have completed my math homework.	2.88	2.92
Ask for help when I do not understand.	2.87	3.24
Memorise things by reading aloud, copying, underlining, highlighting, etc.	2.78	2.95
Do a lot of practice exercises.	2.70	3.05
Check my answers against those given in the books.	2.66	2.90
Work through past examination papers.	2.65	3.11
Study math together with friends, buddies.	2.60	2.99
Copy things from the white board, PowerPoint slides, OHP.	2.53	2.76
Revise my notes regularly.	2.45	2.94
Volunteer to answer questions in class.	2.32	2.70
Do extra math problems on my own.	2.32	2.86
Work on math projects given by the teacher.	2.30	2.66
*Explain math to others, e.g., friends, teacher.	2.29	2.70
Do practical activities, e.g., protractor, abacus, algebra tiles.	2.23	2.59
Do the same problem in different ways.	2.20	2.63
*Keep a notebook of my math mistakes and the corrections.	2.18	2.50
Copy someone's else work	2.12	2.02
Read two or more different math books.	2.12	2.61
*Plan my own schedule of studying math.	2.10	2.56
Do math activities outside class, e.g., puzzles, math trails, math corners at Science Centre.	2.08	2.63
*Make my own notes.	2.03	2.48
Make up my own questions and solve them.	1.99	2.38
Present my solution in front of the class.	1.92	2.42
*Write reflection about lessons, e.g., in a journal.	1.86	2.15
Read about life of mathematicians or history of math.	1.64	2.16
Use software, such as *Excel, Graphmatica,* etc.	1.63	2.10
Search for math resources on the Internet.	1.47	1.95

The most frequently used study behaviour by the polytechnic students was to "work through past year tests and examination papers in mathematics," also perceived to be the most helpful. This study behaviour is also evident at school levels; teachers often design test questions that closely match those from past public examinations, and assessment books that model after these examinations enjoy brisk sales.

Of special interest to the discussion here are some findings about the metacognitive items. On one hand, the three top items in terms of frequency and helpfulness were:

- Recognised when to use the appropriate mathematical formula.
- Checked my work while doing it.
- Analysed mistakes made in solving tutorial problems.

On the other hand, the three bottom items were:

- Asked myself a question about the Maths problem/ topic.
- Planned the strategies before beginning to solve a problem.
- Tried out other possible solutions to a same problem.

A major contrast between these two clusters is that the less frequently used activities require stronger metacognitive skills and greater initiatives from the students than the more frequently used ones. These are characteristics of deep learners. The next section proposes one way to deal with this.

2.4 *Promoting alternative learning strategies*

It is likely that many students do not know about alternative learning strategies and how to use them to study mathematics. Hence, the first step is to enhance their metacognitive knowledge. The following teaching sequence is helpful.

a) Decide on the learning strategy to teach, e.g., concept mapping to link ideas, goal setting to set direction for study, and student question cards and question times to encourage active participation in lessons.

b) Model how to use the strategy by using the "I do – We do – You do" process, thinking aloud, citing cases, etc.
c) Encourage students to practise the strategy on their own and write their own reflection.
d) Ask students to design posters about the strategy and display them around the class. This capitalises on incidental learning.
e) After the students have some practice with the strategy, conduct a mini-debate about the strategy, a technique suggested by Bell (1997). Consider the strategy: *To use a notebook of mistakes for practice and revision is a waste of time*. Divide the class into three groups: one group to support the statement, one group to oppose it, and one group to ask questions and to give the final vote. It raises awareness about effective learning.
f) Guide the students through the four aspects of metacognition covered in Table 8.1. They will take time to internalise a new learning strategy.
g) Provide encouragement when students have demonstrated some successful use of the strategy.
h) After several strategies have been learned, compare and contrast how these strategies can be used in a strategic way. This results in strong self-regulated learning.

As noted above, the scant interest shown by mathematics educators and teachers about learning strategies may have contributed to the unsatisfactory situation in which students can think of only conventional study methods. It is hoped that the above discussion about metacognitive framework, student question cards, research studies, and teaching sequence will stimulate greater interest among teachers and educators to develop innovative programmes to help students become more proactive learners.

3 Metacognition during Problem Solving

The metacognitive aspect of "monitoring of one's own thinking" applies to problem solving. It can be linked to Polya's problem solving model. Implementing this curriculum goal presents the greatest instructional challenge to many Singapore teachers, in comparison with their stronger ability to deal with concepts, skills, processes, and attitudes. There are three reasons to account for this challenge. First, metacognition is a relatively new and abstract construct in education. Its meanings are not well understood. Second, many teachers do not have the personal experiences of engaging in metacognitive activities as past learners of mathematics to make sense of the training they may have received about the theories and praxis of metacognition. Third, teacher training about metacognition is woefully inadequate. For example, in the TEDS-M study (Tatto et al., 2012), the questionnaires used to collect information about teacher education programmes did not include metacognition.

3.1 *Cognition vs. metacognition*

As noted earlier, cognition is about thinking. In mathematics, cognition refers to inductive and deductive reasoning, processing information about concepts and skills in specific topics, analysing data, drawing graphs and diagrams, posing questions and trying to solve them, and other thinking processes. With practice, a high proportion of students can apply these cognitive processes almost automatically and unconsciously to solve routine problems (Clark, 2006). Metacognition, however, means going beyond this automatised thinking by monitoring the use of the cognitive processes. The relationship between cognition and metacognition during problem solving can be shown in Figure 8.2. This model differs slightly from the one given in an earlier chapter (Wong, 2013b) in one aspect: reflection is now included under metacognition, whereas it was absent from the earlier model. Retrospective reflections of learning and problem solving activities are now popular *exit tasks* in lessons, and they provide valuable practice to help students become more metacognitive.

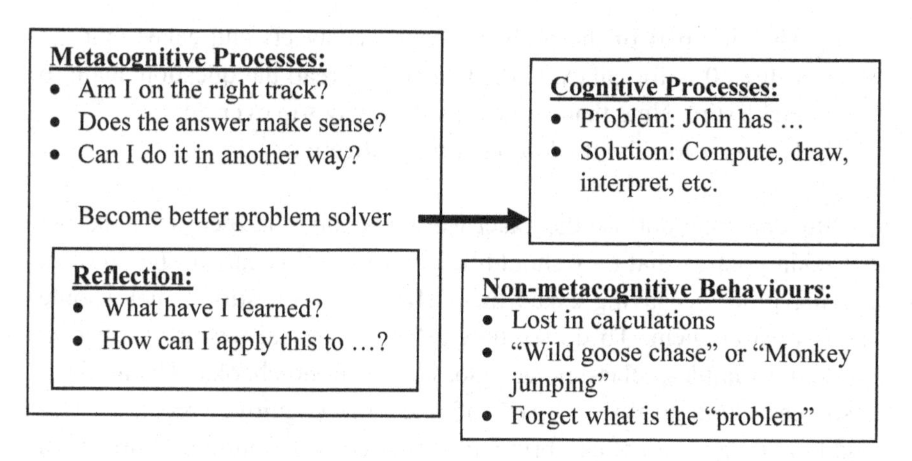

Figure 8.2. Metacognition vs. cognition during problem solving

The distinction between cognition and metacognition is illustrated with the following problem.

> Mary has some sweets and cakes in the ratio 3:5. If there are 30 cakes, how many sweets and cakes does Mary have?

Cognition. The following thinking may be involved:

a) Meaning of ratio and its symbol.
b) Draw a model to show the relationship.
c) Align the number 30 to the correct item in the ratio.
d) Compute the answer using rule about proportion.

Metacognition. Metacognitive prompts are used to raise students' awareness of their thinking and problem solving process. Students may use these prompts in their self-talk, noted by Plato as cited in the chapter's epigraph. The metacognitive prompts that correspond to the cognitive steps in the above problem are as follows:

a) Do I remember the meaning of ratio? If not, how do I find out more about this?
b) Is the diagram complete? Have I included all the information in the diagram?

c) The first part of the problem mentions sweets and cakes, but the value 30 is for cakes. Perhaps, I should read the question again to make sure. Note that careless readers may trip over this.
d) Is proportion about addition or multiplication?

Students may not use the exact wordings above in their self-talk, but the main point is that they should be mindful of this part of the problem solving process. Having asked themselves these questions, they should try to *answer* them. To do so, they refer to external resources, such as worked examples, class notes, glossary, and notebooks. Using these resources strategically is part of the metacognitive process. This additional engagement can bring about the correct solution. However, if the problem is routine and well-practised, metacognitive prompts are not necessary; indeed, insistence on using these prompts in this situation can be counter-productive.

The above framework and example also highlight that metacognition builds on the relevant and requisite cognitive processes. When the prior concepts, skills, heuristics, reasoning competence are lacking or faulty, metacognitive processes are unlikely to lead to the correct solutions. Furthermore, metacognition is at a more abstract and deeper level than cognition, so it is harder to develop metacognition than the underlying cognition. These two observations suggest that students must first master the cognitive components of mathematics before they can develop and apply metacognition. This may be the reason for placing metacognition at the upper end of the Singapore mathematics curriculum framework.

3.2 *Non-metacognitive behaviours*

In the above example, students who do not engage in metacognitive self-talk may end up in unproductive non-metacognitive behaviours. These behaviours are like chasing a wild goose or pursuing a hopping monkey. The following description of these behaviours, written by me in 1991 and then slightly edited in Wong (2002a), still rings true nowadays:

> When solving standard mathematics problems, students normally recall and apply learned procedures in a straightforward way. However, if the problem is unfamiliar, some students simply pick a method and keep persistently on the same track for a long time without getting anywhere. Schoenfeld (1987) described this behaviour as chasing the wild mathematical goose. A different behaviour is also observed: some students jump from one rule to another in a haphazard way hoping to find the correct answer, become agitated, frustrated and finally give up. Teachers who observe both types of unsuccessful behaviours may think that the students have not mastered the skills, and then proceed to re-teach the skills. This may work for some but not for those who have the skills but cannot decide which skills to use and to monitor their use of the skills. A more "metacognitive" teacher believes that the students' difficulty may not be with the skills, rather it might indicate a lacking in self-regulation of the problem solving process. To help the students to avoid such a wild chase, the teacher may ask them to estimate the answer before they become bogged down by rules and computations. (pp. 1-2)

The relevance of making an estimate or informed guess about the answer is a powerful heuristic, which Polya (1969) highly recommended to be used at the initial stage of problem solving. This practice is likely to avert the wild goose chase or hopping monkey pursuit. But Passmore (2007) cautioned that the students "do not get lost forever in pursuing their wrong guesses" (p. 48). The use of metacognitive prompts at different stages of problem solving can raise this awareness and thereby prevent students from getting stuck for too long in trying to get to the expected answers.

3.3 *Developing metacognition*

In this section, I recommend the use of two parallel sets of prompts. One set is used by the teacher during instruction (stressing *you*) and the other by the students as self-talk (stressing *I*). One pair is shown below.

- *Teacher prompt*: When *you* solve a problem, *you* must try to understand what the key symbols mean.
- *Student prompt*: When *I* solve a problem, *I* must try to understand what the key symbols mean.

The teacher prompts function as a scaffold to help students internalise the problem solving process by reminding themselves to use the student prompts. Samples of pairs of prompts for each of Polya's problem solving stages are given in Table 8.5. The prompts can be printed on cards, one card per stage. The students should prepare their own set of cards so that they can refer to them during the lessons and when they do their homework.

Table 8.5

Metacognitive teacher prompts and student prompts for Polya's stages

Polya's Stages	Teacher Prompts	Student Prompts
Understand the problem.	• Do you understand the key words and symbols? • What are you supposed to find?	• Do I understand the key words and symbols? • What am I supposed to find?
Devise a plan.	• Can you think of a similar problem? • Can you draw a diagram?	• Can I think of a similar problem? • Can I draw a diagram?
Carry out the plan.	• Do you need to read the question again? • Do you and your friends use the same method?	• Do I need to read the question again? • Do my friends and I use the same method?
Look back.	• Do you think this is right? How do you know? • Do you need to practice more?	• Do I think this is right? How do I know? • Do I need to practice more?

The following recommendation combines the "I do – We do – You do" teaching sequence with the transition from teacher's public instruction to students' private talk; see Chapter 3, Section 6.2.

a) Begin with the prompts for the *Understand* stage.
b) *I do.* The teacher models how to use the *Understand* prompts to solve a non-routine problem. Both the teacher and students should read out their respective prompts; this is the public talk.
c) *We do.* Use discussion to guide students to work on another problem. The students will read out the relevant prompts and emulate the ways the teacher has used them to solve the first problem. Students also benefit from listening to how other students answer these prompts for the second problem.
d) *You do.* For the third problem, students read out the prompts softly to themselves while the teacher walks around to monitor this.
e) Assign other problems as homework and explain to the students that they should try to whisper the prompts to themselves when they solve the problems. This leads to private talk.
f) Focus on the *Understand* stage for a few lessons until the students become familiar with the technique.
g) Repeat the above cycle for the other three stages. This will take a few more lessons to complete. By now the students should be familiar with the prompts and their use. Some teachers like to cover the prompts for all the four stages during the lesson when they first teach this metacognitive technique. This is quite overwhelming and should be avoided.
h) A potent *I do* attempt is for the class to pose a problem unfamiliar to the teacher for him/her to solve on the spot. The teacher thinks aloud using the prompts so that the students can follow the teacher's thinking, which could be quite messy or unsuccessful. If the teacher becomes stuck, he/she can demonstrate how to use the prompts to get unstuck (see Wong, 2013b, for ideas to get unstuck). Bringing the problem solving process into the open in this way (to make it *visible*) benefits not only the students but also the teacher in helping all of them to be more aware of the functions of metacognition. However, this is quite unnerving for teachers who are mathematically challenged.

i) As the teacher and students master this technique, they can add new prompts to the cards, including prompts for attitudes, motivation, and beliefs about mathematics learning. Ideas for prompts can be culled from the literature; e.g., Di Teodoro et al. (2011); Elder and Paul (2005); Holton and Clarke (2006); Mevarech and Fridkin (2006); Toh et al. (2011).
j) Hold sharing sessions when the students discuss their experiences of using these prompts. Then they write journal entries about this sharing.
k) Encourage students to continue to use this technique in subsequent lessons when they solve difficult problems.

The above approach assumes that the teachers are able to apply these metacognitive prompts to solve mathematics problems and have the competency to teach it to their students, following the broad sequence given above. For teachers who lack both competencies, they should receive adequate training in this area.

3.4 *Investigating metacognition*

Metacognitive monitoring and self-talk cannot be accessed directly because they are mental activities. Researchers of metacognition have used different techniques to gain glimpses of these activities, and each technique has its strengths and weaknesses. By becoming aware of these, one will be more careful in accepting and generalising from reported findings. Brief comments are given below about two commonly used techniques.

A common data collection technique is to ask students to think aloud when they solve problems, and these utterances are recorded and transcribed for analysis. Having to think aloud splits student's attention into several areas: decide on the rule to use, carry out the rule, monitor the progress, remember to talk aloud, and so forth. Many students are not able to do these simultaneously, so the data are incomplete. Furthermore, when the mathematical rules can be applied automatically to solve routine problems, metacognitive monitoring is absent. Hence, when students fall silent, it is difficult to tell whether they are engaged in automatic cognitive processing or they forget to think aloud.

Another popular technique is based on retrospective reflection by the students at the end of a problem solving session. The reflection may be open-ended as in journal writing or structured using a questionnaire or checklist. This technique tends to capture what happens towards the end of the session rather than at the beginning because of the recency effect.

One well-known questionnaire used to study mathematics problem solving in this way was designed by Callahan and Garafalo (1987). They asked students to write answers to questions such as: "Think of everything you do when you practice solving mathematics problems. Why do you do these things?" (p. 22). This technique was used in the *Developing the Repertoire of Heuristics for Mathematical Problem Solving* (MPS) project (Teong et al., 2009). However, most of the comments were very brief, making it difficult to decipher their meanings. There are several reasons for this:

- They did not engage in metacognitive monitoring because of automaticity of the cognitive processes.
- They were not conscious of their cognitive and metacognitive processes.
- They could not express their cognitive and metacognitive processes in writing due to poor English.
- They were not motivated to write extensively.

Despite these concerns, some insights can still be gained. About two thirds of the students attributed their mistakes to carelessness. Some of them mentioned that they could avoid this by writing more slowly or avoid using mental calculation. About half of them believed that their failure to solve a problem was because they had not done it before. About 60% wrote that more practice would improve their problem solving performance; for example, "let us practice more challenging word problems of different types." These comments about practice were so brief that it is not clear whether or not the students were aware of the kind of deliberate practice likely to be helpful.

This brief survey does not do justice to the extensive work done on methodology used to study metacognition. Nevertheless, it is abundantly clear that metacognition is very complex field of praxis and research about student learning.

4 Be Mindful

Metacognitive awareness is akin to being mindful. In recent years, mindfulness programmes are becoming popular in some countries such as the US (e.g., Buoncristiani & Buoncristiani, 2012; Rechtschaffen, 2014) and through the Internet[58]. These programmes train students to be more mindful about their classroom environments and personal emotions and thinking. As a consequence of this training, they listen more attentively during lessons, concentrate on the tasks on hand for longer periods and with less distraction, and develop strong capacity for self-reflection. This will ameliorate the wild goose chase or the hopping monkey pursuit of non-metacognitive behaviours. Zenner, Herrnleben-Kurz, and Walach (2014) reviewed 24 studies about mindfulness programmes and found that the effect size of mindful training on improving cognitive performance was strong (about 0.80). They concluded that mindfulness interventions, though still nascent, hold promise for improving learning.

The basic procedure is to begin each lesson by letting all students spend a few minutes in silence or observe their breathing as in traditional meditation; see a sample lesson at the *Mindful Schools* website[59]. This mindful exercise can also be done when the students are distracted, tired, or restless. By calming their mind, the students are more likely to be productive in completing the cognitive and metacognitive activities.

Implementing mindfulness interventions in schools is, however, challenging. Teachers rarely receive mindfulness training as part of their pre-service and in-service education. Those few who have been trained need to hone their own practice before they can guide their students to be mindful during lessons. Some stakeholders may mistakenly link mindfulness and meditation, just like Yoga, to religious practices that

---

[58] http://www.meditationinschools.org/resources/

[59] http://www.mindfulschools.org/resources/sample-lesson/

conflict with their own beliefs. These issues need to be resolved before this approach is taken seriously as an avenue to inculcate metacognition in mathematics instruction.

5 Concluding Remarks

This chapter elaborates on two aspects of metacognition included in the Singapore mathematics curriculum. The first aspect is self-regulated learning, a neglected area in mathematics instruction and research. The few studies my colleagues and I had conducted in Australia, Brunei Darussalam, and Singapore show that mathematics students in these countries tend to rely on traditional, generic learning strategies, with moderate emphasis on memorisation and rule mastery, and limited attention to active, mathematically focused strategies. To rectify this imbalance, students should learn more about alternative mathematics-based strategies, such as how to ask meaningful questions using student question cards.

The second aspect is about engaging in productive metacognitive behaviours during problem solving. This can be achieved by training students to use metacognitive prompts in self-talk at every stage of Polya's model of problem solving. This metacognitive processing requires that students have mastered the requisite concepts, skills, and processes, so that these resources can be strategically selected and used to solve unfamiliar problems.

This chapter proposes a framework consisting of metacognitive knowledge, metacognitive skills, metacognitive monitoring, and metacognitive reflection. Teachers can conceptualise metacognition during learning and problem solving along these four phases and guide students to make progress from one phase to the other using the teaching sequences discussed above. Since research and theories support the essential role of metacognition in student learning, teachers ought to include metacognition as part of their lesson plans. This is the focus of the next chapter.

Chapter 9

School Curriculum: Prepare Thoughtful Plans

School-based mathematics curriculum consists of two major components: a scheme of work covering daily lesson plans and special curriculum projects. This chapter examines how to thoughtfully design effective mathematics lesson plans and practical worksheets to provide enriched learning experiences.

By failing to prepare, you are preparing to fail. Benjamin Franklin

1 Benefits of Preparing Daily Lesson Plans

Teachers implement the intended national curriculum through the daily lessons they teach. To plan thoughtful lessons, they may find the strategies in Chapters 2 to 8 informative and practicable.

Well-written lesson plans should be complete with the key mathematics contents, worked examples, explanations, questions to ask, practice problems, and homework. Having written down a good plan, the teachers should rehearse and memorise the key examples and questions, so that during the actual lessons, they are able to focus on other important classroom events such as noticing students who are bored or lost, giving appropriate answers to students' questions, maintaining smooth transition from one phase of the lesson to the next, and providing additional help to those who need it, without having to frantically recall what examples or questions to use next. The ability to attend to these diverse events in a mindful way is called *withitness* (Kounin, 1977), which is an essential characteristics of effective teachers. Having thought through the plan, the teachers can avoid giving inappropriate examples and muddled explanations on the spot, which can confuse the students and lead the lesson away from its main objectives. Indeed, students quickly notice poorly planned lessons and react negatively by creating classroom management problems for the teachers.

Pre-service teachers are required to write detailed lesson plans to demonstrate how well they can apply pedagogical theories to devise their plans, especially for lessons they are going to teach during practicum. Based on my limited contacts with teachers, many of them stop writing lesson plans in detail after graduation, although some of them are required to submit very brief lesson notes to the schools. I believe that every teacher should at least write down the examples they wish to discuss at different phases of the lesson so that they do not flounder helplessly through it.

The above arguments do not imply that teachers must rigidly follow their plans because unexpected circumstances need to be handled flexibly on the spot with mindfulness. Think ahead about the following circumstances:

- Students do not have the expected prior knowledge; some revision needs to be carried out and simpler examples given.
- Students make unforeseen mistakes; more examples need to be given on the spot; alternatively, the teachers can defer discussing these mistakes to later lessons after they have given careful thought on these misconceptions.
- Equipment, especially computer, does not always work as planned; the teacher should be able to use different teaching aids.
- Poor time allocation results in completing the work too quickly; the teacher needs to have a collection of interesting mathematics problems that can be readily recalled to fill in the remaining time.
- Poor time management can also result in not being able to cover all the work as planned; those who are mindful of the time can decide on an appropriate place to stop.

It is true that thoughtful plans do not always lead to successful student learning, but hastily prepared plan or lack of planning often ends in poorly-executed lessons with disastrous consequences for student learning.

There are many formats for lesson plans (e.g., Lee N.H., 2009), and most of them include the following essential components:

a) Learning objectives
b) Prerequisites
c) Motivation for the lesson
d) Development and Consolidation
e) Closure

2 Learning or Instructional Objectives

Learning objectives (*learning outcomes*) are targets for student learning, while *instructional objectives (specific instructional objectives, SIOs*) are the purposes of teacher teaching. This distinction is useful because the primary focus of every lesson should be about helping students learn rather than about describing what the teacher intends to do during the lesson. This is highlighted by noting that *teaching and learning are not the same*. In practice, both terms (*learning* objectives and *instructional* objectives) have been used interchangeably because learning and instruction are so inevitably intertwined that it is not very helpful to make fine distinctions between them. In this chapter, the two terms will be used interchangeably.

2.1 *Specific instructional objectives and advance organisers*

Specific instructional objectives (SIOs) are based on behaviourism which stresses that objectives must be measureable or observable. Hence, only action verbs that satisfy this criterion are used in SIOs. These action verbs can be found in Bloom's original and revised taxonomies (Anderson & Krathwohl, 2001; see Fong, 1986) covering the cognitive, affective, and psychomotor domains. Abridged lists are available in various websites [60]. These action verbs are either generic or mathematical. Examples are given below:

[60] www.apu.edu/live_data/files/333/blooms_taxonomy_action_verbs.pdf

- *Generic:* recall, explain, compare, analyse, summarise, critique, relate.
- *Mathematical:* calculate, draw, measure, prove, deduce, estimate, evaluate, construct, factorise.

The *ABCD* model[61] of stating SIOs consists of four parts:
- *A*: Audience, the target students.
- *B*: Behaviours that are observable or measurable, i.e., the action verbs.
- *C*: Conditions under which the behaviours are to be performed.
- *D*: Degree (*depth, standard*) of the competency that students are expected to demonstrate, also called the *success criteria* or *benchmarks.*

Below is an example of an elaborate SIO written using this model:
At the end of the lesson, students will be able to solve at least three out of five percentage word problems correctly within 10 minutes.

Most teachers focus on only part *B,* but it is important to take note of part *D* as well so that the expected performance can be used to judge the effectiveness of the lesson.

A recent trend is to move away from behaviourally stated objectives and to express learning outcomes in broad terms, such as to *understand* mathematics concepts or to *learn* about applications of mathematics. One reason for this change is to focus on the *big ideas* about a topic (e.g., Wiggins & McTighe, 2011), but this raises the issue of the validity of assessment. For mathematics, many teachers like to use *understanding* as an objective, but there are at least five types of mathematical understanding: *conceptual, instrumental, relational, formal,* and *logical* (e.g., Skemp, 1976, 1979). Hence, if this objective were to be used, the type of understanding must be explicitly stated. For example,

> At the end of the lesson, students are able to demonstrate *conceptual* understanding of proper fractions by drawing fraction diagrams.

---

[61] www.mdfaconline.org/presentations/ABCDmodel.doc

SIOs can be written to cover the five factors and problem solving stated in the Singapore mathematics curriculum. Consider the SIOs below for a lesson on equivalent ratios involving natural numbers. Note that the factors inside the parentheses are not part of the SIOs; they are included here to show how each SIO is related to the factors in the Singapore mathematics curriculum.

At the end of the lesson, students are able to:

- *explain* in words the meaning of equivalent ratios (*Concept*);
- *simplify* a ratio by division (*Skill*);
- *show* on a diagram why two ratios are equivalent, say 2:5 and 6:15 (*Reasoning*);
- *solve* word problems involving equivalent ratios (*Problem solving*);
- *give* a real-life example where equivalent ratios are used (*Applications*);
- *demonstrate* their confidence in solving word problems involving equivalent ratios by volunteering to answer questions (*Attitude*);
- *use* metacognitive prompts to check their intermediate steps (*Metacognition*); and
- *work well* with members in a team (*Behaviour; this is not part of the curriculum framework but it is an important non-mathematical objective.*).

Although a comprehensive list like the above example will ensure that cognitive, metacognitive, and affective objectives are included, it is not feasible to include all these objectives in a single lesson. In practice, about three objectives are sufficient for one lesson, and several lessons are required to cover the complete set of objectives.

Since writing SIOs based on the Singapore mathematics curriculum is relatively new to many teachers, more samples are given below.

Concepts. Focus on meanings, representations, and examples.

- Explain the meaning of regular polygon to the class.
- Give two examples of a linear function.
- Identify congruent triangles from a set of given triangles.
- Describe complementary angles in words and diagrams.

Skills. State the procedures and rules in specific terms.

- Calculate the average speed of a journey described by a speed-time graph.
- Given the lengths of two sides of a right-angled triangle, determine the length of the third side.
- Use the Sine Rule to find the missing sides and angles of a triangle.
- Construct two similar triangles accurately using ruler, protractor, and compasses.

Processes. This covers reasoning, communication, modelling, etc.

- Prove the Sine Rule.
- Write down the instructions to be used to estimate the height of the school's flagpole using trigonometry.
- Use the sine function to model the number of hours of daylight over a given year.

Attitudes. These objectives may be assessed through observations, listening to students at work, and journal writing.

- Express interest in solving compound interest problems by sharing reading about this topic from newspaper reports.
- Demonstrate perseverance by working on the given problem for at least 20 minutes despite being stuck.
- Show engagement by volunteering to solve a problem on the board.

Metacognition. Check these outcomes through discussion, journal writing, and checklist.

- Describe the thinking process when solving the given problem.
- Explain when to use the Sine Rule or the Cosine Rule to find angles of a triangle.
- Keep records of errors and the corrections.

Advance(d) organisers (Ausubel, 1968) can be used to state the objectives of a lesson. These advance organisers (AOs) specify the mathematics contents to be learned in the lessons. They include definitions, formulae, or principles. They help to link the new contents to previous learning. The links can be shown using concept maps; see Figure 9.1. This concept map may be shared with the students at the beginning of the topic.

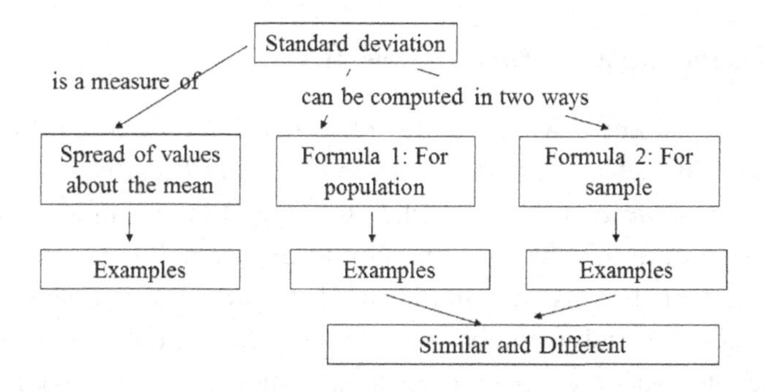

Figure 9.1. Advance organiser about standard deviation using concept map

SIOs and AOs differ with respect to the use of action verbs and mathematics contents. This difference is illustrate below.

- *SIO*. Able to multiply integers. (The students know the expectation but do not know the rule at this stage.)
- *AO*. The product of two integers of the same sign is positive. (The mathematics rule is stated, but the students do not know what they are supposed to do with it because there is no action verb.)

It is possible to mention both SIO and AO, but this is overdoing the function of objectives:

- *SIO* and *AO*. Able to multiply integers by using the rule that the product of two integers of the same sign is positive.

AO is effective in direct instruction because the teacher is basically following the 3-step method below, frequently used in public speech[62]:

- Tell them what you are going to tell them.
- Tell them.
- Tell them what you have told them.

---

[62] http://www.businesstown.com/presentations/present-tell.asp

A disadvantage of stating AOs at the beginning of a lesson is that doing so removes the *surprise* element, which is fun for inquiry-based lessons. Whether SIO or AO is used, it must be about the *new* learning to be covered in that lesson.

2.2 *Informing students about SIOs or AOs*

Students should know where the lesson is heading by being informed of the SIOs or AOs. These should be stated in student-friendly terms instead of the *educationese* used in teacher training. For example, "to relate factorisation of an algebraic expression to expansion" is not meaningful to most students because it is not clear what *relate* means in this context; a more student-friendly version is "to check factorisation by expansion."

Some educators recommend replacing "students will be able to" by "I can" when these objectives are shared with the class to stress that students are the main audience of lessons.

The following points show how to use SIOs or AOs in lessons.

- At the beginning of the lesson, state: *In the last lesson you learned about ___; today you will learn ___; this is important because ___ if you understand the lesson, you will be able to ___.*
- Ask students to explain what they understand by some of the key words in the SIOs or AOs; the teacher may highlight new terms that will be addressed during the lessons.
- Place these objectives at a prominent part of the whiteboard and tick off the items when they have been covered; students will then be able to monitor where the lesson is progressing. Teachers may also ask the class, "Where are we now?" to promote such monitoring.
- If the lesson covers several SIOs, talk about them one or two at a time, instead of showing all of them at the beginning of the lesson because this is quite daunting.
- At the end of the lesson, discuss with the class which objectives have been dealt with adequately and which have not and to be covered in future lessons.

- State the objectives for practice problems given in the worksheets and homework to help students appreciate the purpose for doing the problems.
- It is not effective to show or read off the objectives on PowerPoint slide at the beginning of the lesson, which is frequently done by ICT-savvy teachers. The reason is that the list will soon disappear from the screen and it is not possible to use it to check on the progress of the lesson, as noted above.

3 Assess and Activate Prerequisites

Students bring to lessons a wide variety of prerequisites in the form of concepts, skills, reasoning ability, study habits, beliefs, attitudes, and life experiences. These assorted prerequisites affect how well they learn from the present lesson.

For cognition, meaningful learning of new contents builds on prior knowledge, and this was asserted by Ausubel (1968) in oft-quoted statement below:

> If I had to reduce all of educational psychology to one principle, I would say this: The most important single factor influencing learning is what the learner already knows. Ascertain this and teach him [*sic*] accordingly. (frontispiece)

When student achievement is predicted using student characteristics, prior knowledge is the strongest predictor with effect size of about 1 for many school subjects (Hattie, 2009).

Having decided on the learning objectives, the teacher then determines ways to assess the prerequisite knowledge students need to have and to activate it so that they are ready for the lesson. This has to be a conscious act on the part of the teacher, because, according to Pinker (2014), "When you've learned something so well that you forget that other people may not know it, you also forget to check whether they know it" (p. 63). Checking that students actually know what they need to know involves more than asking the class in a perfunctory way whether they have any query about the previous work.

Prerequisite knowledge for a particular topic can be inferred from curriculum documents, textbooks, teacher guides, personal experiences, task analysis, and the contents of earlier lessons in similar and different topics. Since there are numerous possible preceding knowledge and skills, only those directly relevant to the present lesson need to be considered. For example, the first lesson on percentages at Grade 5 requires knowledge of fractions, division of whole numbers, and part-whole relationship. It is not necessary to consider addition and subtraction of whole numbers, because these skills should have been mastered at lower grade levels and will not hinder the understanding of percentages. Novice teachers tend to cover very basic knowledge but miss the ones really required for the present lesson.

A typical technique to assess prior knowledge is to give the class a short quiz (or diagnostic test), and this can be done orally or in writing. Then determine the proportion of the students who have this knowledge by choosing from a variety of techniques: show of hands, the thumb up/sideway/down gesture to show "agree, uncertain, disagree," Post-it notes, bingo games, mini-whiteboards, interactive whiteboard, or computer-based clicker tools. Students may work in groups to answer the quiz so that they can help one another in this preview. This not only promotes cooperative learning but also reduces the number of student scripts to collect for checking. A third approach is to analyse in detail the homework of previous lessons so that misconceptions can be discussed with the whole class or individual students. If many students are found to have inadequate prerequisites during the lesson, the teacher needs to provide appropriate review and slow down the pace of the lesson.

To assess relevant prior life experiences or attitudes, ask students to respond to short checklists about these experiences or to share things they have encountered, e.g., reports about earthquakes, advertised discounts during sales, books about mathematicians, and new discovery in mathematics. Questionnaire data can be collected prior to the lesson so that relevant ideas can be exploited as motivators to stimulate student interest in the present topic.

4 Motivation for the Lesson

Chapter 8 has dealt with ways to motivate students to learn mathematics. The activities suggested under the *M_Crest* framework can be selected for this phase of the lesson plan. These are also called *lesson starters*.

The motivational activity should be short, memorable, and relevant to the topic. Some teachers like to show YouTube videos or films that are not related to the mathematics to be taught with the sole purpose of eliciting laughter. This is counter-productive because it conveys the unsavoury message that mathematics is so tedious that one has to be entertained by unrelated matters before working on it. Of course, it is appropriate to use videos that can serve both purposes.

An effective practice is as follows. When the teacher first enters the classroom and before he/she attends to chores such as taking attendance, distributing worksheet, and returning homework, put up the motivator (e.g., a real-life problem) on the board and ask the class to think about it. Make this into a routine by using the same section of the board for every lesson. This helps students to get ready for the lesson, sets the tone, and inculcates good study behaviours. This practice is especially powerful if the teacher refers to the motivator throughout the lesson or leads students to finally solve the problem when they have learned the new contents. The lesson plan will include details of the motivator and how it is to be used in the lesson.

5 Development and Consolidation

The main development of the lesson depends on the pedagogy used, especially the two main types, direct instruction and guided discovery. In either case, the development portion of the lesson plan can be divided into more than one episode. The episodes may involve direct instruction or guided discovery so that the students will experience both types of activities. Some episodes will include seatwork or class practice. There should be logical progression from one episode to another.

For each episode, the teacher should write down in the plan the key questions to ask, the likely wrong answers from the students, and the follow-up responses. Going through these items during planning will alert the teacher to potential obstacles and inadvertent digressions and thereby try to pre-empt them from occurring during the lesson. Three examples are given below.

Example 1. The teacher has explained how to compute discount and posed the following problem. A shop gives a discount of 20% in March. For weekends, a further 50% discount is given.
Key question. What is the final discount for weekends in March?
Likely error. 70%. Students tend to add percentages without paying attention to the bases for the percentages, which are different here. Some students (also adults) have difficulty identifying the base in percentage changes.
Follow-up. Draw a diagram to show the base (100%) for each type of discount. Diagrams are powerful medium for thinking about relationships among numbers.

Example 2. The teacher has explained how to multiply a decimal by 10 and 100.
Key question. When you multiply a natural number like 12 by 10, you add a zero at the end. Can you do the same when you multiply a decimal by 10? This is to probe conceptual understanding of the rule and to help students use compare and contrast in their reasoning.
Likely error. Yes. The same rule should apply whether the number is a natural number or a decimal.
Follow-up. There are several possibilities: illustrate the multiplication using 10×10 grids; compare the difference between 0.12 and 0.120, say; teach estimate skills when numbers are multiplied by powers of 10; use a calculator. When several possibilities exist, arrange them in the preferred order for discussion and indicate this order in the lesson plan.

Example 3. The teacher has explained the properties of similar polygons with some examples.
Key question. Place an A4 and an A5 piece of paper side by side as shown in Figure 9.2(a) (not to scale). Are they similar?

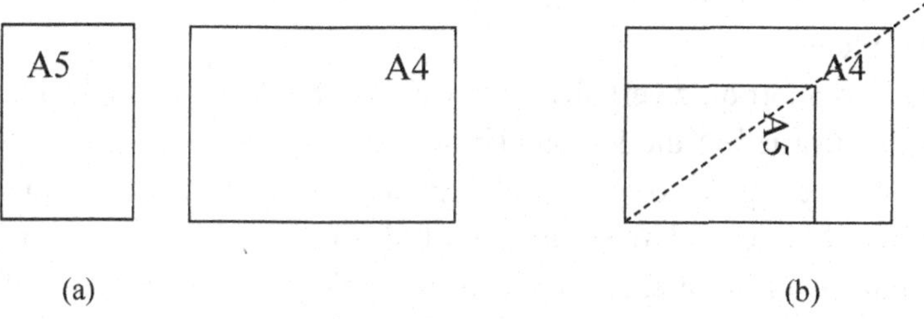

Figure 9.2. Are A4 and A5 paper similar? (not to scale)

Likely error. No because two of the sides are equal in length, hence not proportional. In fact, many teachers in my workshops gave this response! *Follow-up.* Explain that similarity of shapes does not depend on orientation. Let them try again. Eventually some students (and teachers) will arrive at Figure 9.2(b). Alternatively, ask them to measure the sides and check whether the lengths are proportional or not, within measurement errors.

The materials to be used, such as manipulatives, coloured paper, worksheets, software, and Internet websites (double check that they are still available) should be mentioned in the plan and checked prior to the lesson to ensure that they work accordingly.

6 Closure

Every lesson should have a thoughtful closure. Some popular ways to do so are given below[63].

- Refer to the learning objectives and check those that have been adequately covered and what will come next. Say something like this: *Today we did ___, and in the next lesson, we will continue with ___.*
- Summarise the main conclusions if the lesson is based on guided discovery.

[63] See http://k12edresources.com/?p=229 for more examples.

- Compare initial and final responses to the motivator or starter problem.
- Give a short quiz verbally or in writing about the key concepts and rules learned in the lesson. These quizzes should not ask students to derive a proof or give oral explanations to the class, because of limited time and the purpose of the quiz, which is to catch a glimpse of what most students are able to do at the end of the lesson. This complements evidence about student understanding gathered from Q&A interactions and classwork.
- Set reading, a challenging problem, and homework as preparation for the next lesson.
- Ask students to write short reflections in their journals or exit cards. The 3-2-1 approach is quite popular:
  - Three things that you learned in this lesson.
  - Two things that you are not sure about.
  - One thing that surprised you.

  These reflections are collected and read so that interesting points or questions can be discussed in subsequent lessons.

An ineffective closure is to ask the generic "any questions?" question. Many students do not know how to ask specific, meaningful questions in short notice, and most students are unlikely to respond because they may be thinking of the next lesson. Instead, try journal writing.

Pre-service teachers are expected to write plans for closure but few of them actually implement their plans due to poor time management or negligence. These teachers should develop the habit of stopping all activities, including those still in progress, to allow time for closure. Research on teacher effectiveness generally supports strong closure of lessons because of the recency effect, which suggests that closure activities can leave powerful impressions on the students.

7 Overall Organisation

The plan should include the time allocated for each phase of the lesson, and the following is a rough guide for a 60-minute lesson.

a) (5 min) Attendance, settle down; write down objectives and/or motivator.
b) (10 min) Assess and activate prerequisites.
c) (5 min) Motivation for the lesson; refer to the motivator if left on the board throughout the lesson.
d) (35 min) Development and consolidation.
e) (5 min) Closure.

Different education systems have their distinctive ways to organise lessons. The Singapore mathematics curriculum recommends that a typical lesson proceeds in three phases called *readiness, engagement,* and *mastery.* Readers can readily relate these three phases to the five components above.

In England, its national strategies require that every mathematics lesson at primary level comprises three parts: *a whole-class mental mathematics starter, the main teaching activity,* and *a plenary review.* The mental mathematics starter (10 – 15 minutes) is used to develop quick recall of number facts and flexible mental strategies. The main teaching activity (30 – 40 minutes) is a combination of whole-class teaching, group work, and individual practice. During this part of the lesson, the teacher collects evidence to evaluate student progress. The final plenary review (10 – 15 minutes) is the closure. A recent study of 125 Grade 5 classes in England found that plenaries were carried out in half the mathematics classes observed, and even in good schools, these plenaries "were rushed and only covered material at a surface level ... as if the teachers included a plenary because they knew they were supposed to but didn't have enough time either to plan or deliver the plenary properly" (Siraj-Blatchford et al., 201, p. 68). Similar trends may be found in other education systems as well.

A different lesson organisation is worth pursuing. This is designed and marketed by *Scholastic* for its *Math180* [64] programme. Its instructional model utilises blended learning in a specially designed classroom environment. The lessons are carefully scripted. Each lesson of about an hour long begins with whole-class teaching of about 10 minutes, and this includes motivating a new topic with real-life applications, making connections to prior topics, and developing mathematical practices. At the end of this short session, the students are split into two groups, called *group instruction* and *personalised software.* Students under *group instruction* complete further activities about the topic just taught facilitated by the teacher, while those in the *personalised software* group carry out individualised learning using pre-designed computer-based materials. The two groups switch activities after about 20 minutes, but the teacher stays with the *group instruction* group all the time. Note that there is no plenary closure for the whole class. After class time, the students complete a personalised playlist of games and activities under its *Brain Arcade.* Student work is captured online and teacher has access to extensive data analytics to monitor the progress of individual students and the whole class. The purpose of describing a commercial package here is to show that instructional innovations can be found outside official agencies and research projects. It also exhorts teachers to devise their own lesson organisation to suit their teaching preferences and the needs of their students.

8 Review of Lessons

After every lesson, the reflective teachers will analyse it by considering the following aspects:

- What went well; has student's learning improved? State the evidence and avoid vague conclusions such as "The lesson was a success."
- What did not go so well? Examine the evidence.
- Action; what to do next?

---

[64] http://teacher.scholastic.com/products/math180/instructional-design.htm

- Support the above reflection with relevant pedagogical principles. This stimulates the theory-practice linkage underpinning teacher pedagogical knowledge.

The following example has included the above four aspects:
I used fraction cards to teach about fraction addition … The lesson was successful because 90% of the students could do problems 1 to 3 correctly … The fraction cards embody the concept of addition in a fun way … (*reflect on principles*) … The technique did not work well with big fractions such as … I will design a similar set of cards for teaching decimals to establish a link between fraction addition and decimal addition …

On the other hand, the followings are not reflective, although they may be extended by adding some of the aspects mentioned above:

- Describe the activities carried out in the lesson without further comments.
- Paraphrase readings without relating to the lesson.
- Give mathematical solutions.

Much practice is needed to hone the skills to write in-depth reflection, and this can be an important component of teacher professional development.

9 Worksheets

Worksheets are becoming very popular in teaching because teachers can design them by selecting relevant materials from various resources to suit the specific learning needs of their students, rather than being constrained by the textbooks used in their schools. Complex instructions and diagrams can be included in the worksheets to save time copying them on the board. Students also use worksheets for revision. Several websites[65] allow teachers to easily create their own worksheets.

[65] http://aplusmath.com/

Well-designed worksheets should consist of worked examples and exercises that are arranged in order of difficulty to match student ability. These differentiated worksheets can be at three levels: *basic*, *intermediate*, and *advanced*. The basic version has simple problems that most students are able to solve. The intermediate version provides a sequence of key steps leading to the final solution. Finally, the advanced version provides challenges to the more able students.

Some worksheets contain problems that are beyond the standards expected by public examinations. They are included in order to set high standards for school work and to challenge the students. This practice will lead to undesirable stress on the weak and average students. Following from the above suggestion of differentiated worksheets, the problems should be arranged in groups and the expected standard for each group spelt out clearly at the heading of the group. Four groups of problems are as follows:

a) Basic problems that every student must be able to do at the end of the lesson.
b) Problems at the *Paper 1* level or the *Foundation* level (primary schools) or the *Normal* level (secondary schools); readers who are not familiar with the Singapore system may wish to check out the meanings of these labels at the Ministry of Education[66] website and the Singapore Examinations and Assessment Board[67] website.
c) Questions at the *Paper 2* level or the *Standard* level (primary schools) or the *Express* level (secondary schools).
d) Challenge problems. These are beyond the expected standards and are to be attempted only by the truly interested or capable students.

During the lesson, most of the students should work through the problems in the above sequence, but many of them do not need to reach the challenge level. This also educates students to set the benchmarks they wish to achieve, thereby taking greater responsibility for their own learning. This may alleviate unnecessary stress on them.

---

[66] http://www.moe.gov.sg/

[67] https://www.seab.gov.sg/

There are some shortcomings in the use of worksheets. Unmotivated students simply copy things onto the worksheets just to *get through the worksheet* as quickly as possible. They do not fully engage their mind on the learning tasks. Perhaps, many students are not aware of the reasons for doing the worksheets or tasks (Bell, 1997). This can be rectified by attaching learning objectives to the worksheet problems, informing students about the expected standards of the problems, and educating them about desirable study behaviours (see Chapter 8).

It is also necessary to pay close attention to what students are actually writing into the worksheets, especially for the ubiquitous practice of filling in the blanks. Consider the following example:

The product of two integers of the _____ sign is _____.

Students need to write down only two words: *same, positive.* Many of them may not even pay attention to the whole sentence as it is read out by the teacher or by themselves. Students who are weak in oral English will miss it, resulting in no understanding at all. On the other hand, getting students to copy the whole sentence into their notebook allows for fuller processing of the rule. This should be done for simple statements similar to this.

The following three features found in worksheets also deprive students of the important opportunity to generate mathematical objects from scratch on their own. These weaknesses should be avoided.

- Fill in partially completed tables of values; they do not learn to make decisions about the format and size of the tables to suit the nature and amount of data.
- Graphs are already plotted with the proper scales, so students do not acquire the mathematical sense of how scales affect the shapes of the graphs.
- Points are already plotted and the students just write down the coordinates. They do not learn how to plot the points.

10 Unit Plans and Scheme of Work

The unit (or block) plan covers a series of lessons for a particular topic. A collection of unit plans constitutes the scheme of work (SOW) for a term and a whole year. The SOW is usually determined by a team of experienced teachers at the beginning of the school year. A simplified version of the SOW may be given to students and parents to keep them informed of the students' study load.

Learning objectives for the unit plans and SOW should include broader aims that subsume the SIOs of individual lessons. It may take the following form:

At the end of this unit (after 10 lessons, say), you will be able to …

Unit plans and SOW are more than a list of topics to be taught. The following points can be used to prepare SOW at grade level.

1) Consult the curriculum documents to identify contents covered in the grade levels below the current level, including primary school level, if necessary. Use this information to establish linkage of topics across different levels. However, many pupils forget what they have learned in previous years, so include time to activate prerequisites.
2) Construct a concept map or conduct a topic analysis to identify the hierarchy of concepts, skills, and principles involved.
3) Examine the textbooks, teacher guides, workbooks, past examination papers, and other resources for ideas about sequence, scope, and activities. Consult SOW gathered from other schools and commercial publishers.
4) Allocate time for each topic based on past experience. Include time for revision, tests, examinations, and other school functions.
5) Include the schedule for different types of assessment with sample items and marking schemes, if necessary.
6) Consider learning experiences for differentiation and extension to match student ability and needs.
7) Include in the SOW the essential resources to be used.

The final document should be one that teachers can readily use to design lesson plans that are logically sequenced, experientially relevant and rich, and carefully aligned with the intended curriculum.

11 Concluding Remarks

The above discussion stresses that a lesson plan is not just a list of the tasks teachers plan to carry out in their lessons. The plan should be objective-focussed rather than task-based so that the components on prerequisites, motivation, development, consolidation, and closure aim to achieve the learning objectives. Writing out details for every lesson requires considerable effort and time to complete, and this is not practical for most teachers, given their heavy teaching and administrative loads. Despite these constraints, every teacher must prepare the essential examples and questions to use in lessons. At least once a term, they should make the extra effort to write detailed plans so that they can investigate the effectiveness of their plans in order to improve their own teaching, which will impact on student learning. Furthermore, they can share their reflection of the lessons with colleagues to promote professional development. Other ways to help teachers hone their mathematics content knowledge and teaching skills are the subjects of the final chapter of this book.

Chapter 10

Professional Development: Become Metacognitive Teachers

This chapter is about how metacognitive teachers can proactively apply learning principles to deepen and broaden their mathematics knowledge and hone their teaching skills. These areas of teacher professional development are subsumed under *teacher standards.* A framework is proposed to identify the mediating factors that link professional development to student learning outcomes. Five principles of training pedagogy are then enunciated to show how the quality of training activities can be enhanced. Finally, the professional development and practice of mathematics teacher educators themselves are noted.

Stay hungry. Stay foolish. Stewart Brand, cited by Steve Jobs[68]

I am always doing that which I cannot do, in order that I may learn how to do it. Pablo Picasso (1881 – 1973)

1 Need for Professional Development

Much has been written about teachers being the most important factor in student success in learning, so it is important that teachers continue to learn to realise this critical role. They should also remind themselves of their initial professional motivation, other than getting a job, to become teachers and why in mathematics. Many teachers believe in making

---

[68] https://www.youtube.com/watch?v=UF8uR6Z6KLc&t=830

differences to the life of their students by helping them develop their full potential and become good citizens. Similarly, teachers should develop themselves to their full potential as teachers as well as individual citizens. There is, indeed, no end to the lifelong journey to grow from being a pre-service teacher to a masterful one.

Many teachers have already developed teaching habits that *work* and may not be willing to change. Some may not be able to change. Others are not given the opportunities to participate in professional development. These obstacles (not seeing the need to change, unwilling to change, unable to change, lack of opportunities) impede teachers' progression from being just adequate to highly competent during their lifelong career.

Metacognitive teachers, however, are those able to act proactively to overcome these obstacles on their own rather than to wait for opportunities to be offered to them. They follow the path suggested by Picasso to teach in new ways so that they can learn from this novel experience. A metacognitive teacher is more than a reflective one. Metacognitive teachers can use the knowledge and ideas gained from reflection of past lessons to understand areas for improvement, experiment with new techniques, and monitor their professional learning by gathering evidence, in ways parallel to what metacognitive students are able to do.

Teachers' professional learning covers three key areas. First, they need to deepen and broaden their mathematical content knowledge. This includes mastering the fundamentals of the mathematics they are teaching (Ma, 1999) and learning new mathematics. Second, they need to hone their teaching skills by experimenting with new techniques. Third, they need to thrive in their enthusiasm to teach under different classroom situations and to enjoy or have fun doing so. Teacher enthusiasm is contagious, but it seems to decline as teachers keep repeating similar lessons years after years. Professional development should address this issue as well. The following sections will discuss the first two areas in detail.

2 Deepen and Broaden Mathematics Knowledge

Competent mathematics teachers should continue to deepen and broaden their mathematics knowledge (ACME, 2002). According to Ma (1999), depth of understanding comes from the capacity to connect relevant topics to strengthen conceptual and reasoning power. For example, a teacher can relate the concepts of division across whole numbers, fractions, decimals, integers, algebraic expressions, and matrices (in terms of multiplicative inverses). It also means going deeper into a topic to understand its mathematical properties beyond routine procedures. This is called *specialised content knowledge, content knowledge for teaching,* or *mathematics knowledge for teaching* (MKT) (Ball, Thames, & Phelps, 2008; Morris, Hiebert, & Spitzer, 2009). The term *horizon knowledge* is used to cover "an awareness of how mathematical topics are related over the span of mathematics included in the curriculum" (Ball, Thames, & Phelps, 2008, p. 403).

A widely discussed example of MKT is to know why one converts fraction division into multiplication by inverting or flipping the divisor. Another example is to appreciate the concept of place value by studying numeration systems that use different bases, such as the binary, octal, and hexadecimal systems, as well as systems that do not use place value, in particular, the Roman numerals. The purpose of studying these numeration systems is to explore how rules of operations can be developed in mathematically consistent way within each numeration system and not to master these rules per se. Other examples of which teachers need to have fundamental understanding include *why 1 is neither prime nor composite* and the ambiguity about certain mathematical objects such as 0^0.

There is evidence that pre-service teachers, even the high achieving ones, "often miss many of the important mathematical connections in school mathematics and frequently exhibit their own misconceptions" (Cooney, 2001, p. 27). Practising teachers have been found to exhibit similar superficial understanding of essential school mathematics. To address this, mathematics teacher education programmes in many countries include *subject knowledge courses* to enable pre-service teachers to study school mathematics at advanced level (e.g., Ball,

Lubienski, & Mewborn, 2001; Lim-Teo, 2009; Usiskin, 2001). Several projects attempt to measure this type of knowledge, e.g., *Learning Mathematics for Teaching* (LMT)[69] (Ball, Thames, & Phelps, 2008) and the TED-M Study (Tatto et al., 2012). Felmer and his Chilean team (2014) have produced a comprehensive set of indicators of various types of mathematics content knowledge for primary school teachers. The metacognitive teachers can check their understanding against these tests and indicators and take the necessary proactive steps to close gaps in their mathematics knowledge.

Breadth of mathematics refers to knowledge about a wide range of topics, including those not in the official intended curriculum. By studying these topics at conceptual but not necessarily rigorous levels, teachers gain a broad perspective of different branches of mathematics and keep abreast of new mathematics developments, such as the largest known prime or that Fermat's Last Theorem and the Four Colour Theorem have been proved. With this knowledge, they can stimulate student interest about mathematical discovery, answer queries about the applications of mathematics in diverse fields, and make lessons more interesting. Books that popularise mathematics are good starting points, and they may inspire teachers to study certain topics to greater depth. A few noteworthy books are: Ayres (2007), Courant and Robbins (1947), Dewdney (1993), Gigerenzer (2002), Kolata and Hoffman (2013), Paulos (1995), and Stewart (2006, 2009, 2011). The mathematics in these books can be adapted as challenging problems for high school students.

The third type of knowledge is academic mathematics, such as Abstract Algebra, Multivariate Calculus, and Set Theory, which are normally studied during undergraduate years. These courses extend technical skills and prepare teachers to teach new contents that may be included in the curriculum in the future. However, studying these academic mathematics courses does not seem to prepare pre-service teachers to have deeper understanding of school mathematics. The reason is that these tertiary courses do not re-visit school mathematics, and students with only instrumental understanding of school mathematics do not have the opportunity at university to bridge the gap in their

---

[69] http://sitemaker.umich.edu/lmt/home

conceptual and relational understanding. For example, although pre-service teachers encounter π in many advanced mathematics courses, many of them still do not know that it is defined as the ratio of the circumference to diameter of a circle, believing instead that it is exactly $^{22}/_{7}$ (Wong, 1997a; see Chapter 4, Section 3.2). Hence, working with a piece of mathematics at an advanced level does not necessarily transfer downward to bridge gaps in understanding of the same knowledge at a lower level. Wong (1997c) referred to this as a lack of *downward transfer,* while Cuoco (2001) called this *vertical disconnect* between tertiary mathematics and school mathematics. Correlational studies show that the number of tertiary mathematics courses taken by the teachers had only weak direct effects on their student learning (Fennema & Franke, 1992). It is not clear whether there is any improvement to this issue in the past twenty years. What is obvious is that mathematics educators tend to focus on MKT instead.

The German mathematics educator Wittmann explained in a recent interview that teachers act as tour guides to their students through the mathematical landscape (Akinwunmi, Höveler, & Schnell, 2014). They need to know the valleys, peaks, scenic spots, country roads and other features to help all students in their hiking, even though they do not expect every student will reach the same peaks. This metaphor should spur all teachers to deepen and broaden their mastery of mathematics. Metacognitive teachers can set their own goals, for example, to study one new mathematics topic per year. They will experience once again the challenges, cognitive dissonance, and joy of mastering new mathematics using the learning approaches discussed in earlier chapters, like looking for examples and non-examples, constructing concept maps, exploring the topics in multiple representations, and solving self-generated problems using heuristics. This invaluable experience enables the teachers to validate the effectiveness of these techniques they use to help their students learn.

3 Master New Pedagogy

Most teacher professional development activities aim to enhance teacher pedagogy. This section covers the *what* of new pedagogy in terms of *teacher standards* and the *how* of achieving this through teacher professional development programmes.

3.1 *Teacher standards*

Pedagogy covers both generic teaching techniques and maths-based methods, and these are grouped under *teacher standards,* which also includes qualities that teachers must demonstrate before they graduate from teacher training and during their service.

The *Standards for Excellence in Teaching Mathematics in Australian Schools*[70] by the Australian Association of Mathematics Teachers is an example of these standards. It covers professional knowledge, professional attributes, and professional practice. The standards describe what excellent mathematics teachers are supposed to be able to accomplish, say, "to initiate purposeful mathematical dialogue with and among students." A more detailed approach is the *Australian Professional Standards for Teachers*[71] by the Australian Institute for Teaching and School Leadership. It spells out seven standards, each one in four different levels: *Graduate, Proficient, Highly Accomplished,* and *Lead.* These levels generally progress from being able to implement specific strategies after graduation, becoming proficient in making the strategies effective for student learning, supporting colleagues in strategy use, and finally leading by demonstrating exemplary practices and reviewing them through collection of data. These standards are captured in the framework shown in Figure 10.1; it is adapted from Wong (2009b) and covers standards for metacognitive mathematics teachers.

There are six types of standards in the framework. The standards on mathematics knowledge include mathematics knowledge for teaching,

---

[70] http://www.aamt.edu.au/Better-teaching/Standards

[71] http://www.aitsl.edu.au/australian-professional-standards-for-teachers/standards/list

horizon knowledge, and academic knowledge. It forms the bedrock of metacognitive practice and professional growth. With this knowledge, the teacher can better organise topics in the mathematics curriculum in terms of coverage, sequence, and pacing, so that they can monitor the mathematical development of the students, address misconceptions, and stimulate creative problem solving.

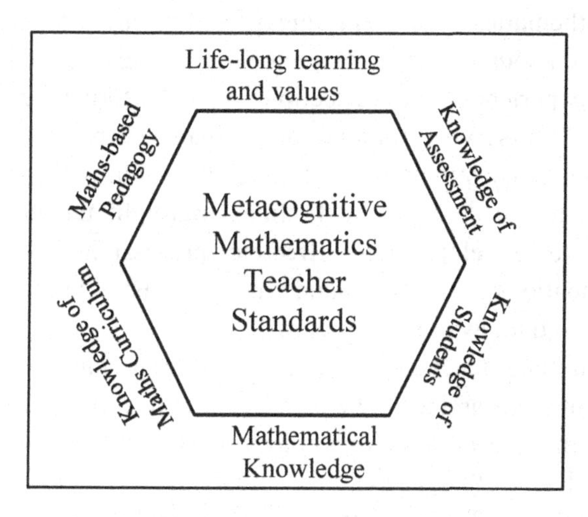

Figure 10.1. Framework for Metacognitive Mathematics Teacher Standards

Research has shown that generic teaching techniques have only weak to moderate effects on basic arithmetic skills (AERA, 2005). Hence, it is more pertinent to develop maths-based pedagogy, which applies generic teaching techniques to match specific features of different mathematics topics. Teachers must also be able to administer different types of assessment to collect data about student learning for formative, summative, and ipsative purposes (see Chapter 1). The metacognitive teachers are aware of their professional motivation and values of becoming a teacher. They are able to determine their current level of competence, say along the four levels discussed above (graduate-proficient-highly accomplished-lead), and plan relevant development activities to rise to the next level. This shows their personal commitment to engage in lifelong learning to benefit their students.

3.2 *Training pedagogy*

Teacher professional development takes many forms: listen to talks, attend workshops, enrol in certified courses, share ideas with colleagues informally, read education materials, engage in lesson study and action research, and participate in professional learning communities (see Chua, 2009 for some Singapore examples.). From a survey of 66 secondary school mathematics teachers in Singapore, Fan and Cheong (2003) found that the main sources for developing pedagogical knowledge were "own teaching experience and reflection" and "informal exchanges with colleagues." In-service training and organised professional activities were of secondary importance. Although the sample was small and taken more than a decade ago, the findings highlight the need for teacher educators to develop more effective professional activities. Many recommendations can be found in the vast literature on teacher education, and a few are cited below.

- "stimulate teachers to re-think, to experiment, to make fresh distinctions and to probe those distinctions to explore how they are informative in enabling choices related to teaching and learning" (NCETM, 2009, p. 14).
- "the most effective forms of professional development seem to be those that focus on clearly articulated priorities, providing ongoing school-based support to classroom teachers, deal with subject matter content as well as suitable instructional strategies and classroom management techniques and create opportunities for teachers to observe, experience and try new teaching methods (OECD, 2005, p. 128).
- "professional development must provide teachers with a way to directly apply what they learn to their teaching" (AERA, 2005, p.2).
- "short courses (half/one/two day) are optimally effective when there is time for teachers to reflect on what has been learnt, to seek the best ways of implementing the ideas and methods in the classroom and to reflect on these practices in an informed way" (ACME, 2002, p. 4)

- Ten principles enunciated by Clarke (1994/2007). Among these principles are: addressing issues raised by the teachers, involving groups of teachers rather than individuals, changing teacher beliefs, and encouraging them to set goals for personal growth. AERA (2005) also emphasised collective participation which tended to create more active learning among the teachers.
- Five characteristics nominated by DeMonte (2013): alignment with systemic goals, modelling of teaching strategies for contents (this is similar to maths-based pedagogy), active learning of these strategies, follow-up feedback, and teacher collaboration.

These recommendations can be summarised into five *training pedagogy,* which can be used in teacher education courses. These principles were originally developed for pre-service training (Wong, 2005), but they apply to in-service courses as well. They are briefly explained below.

- *Spiral and Developmental Principle.* Explain to teachers how to apply the maths-based pedagogy consistently to different topics. At each re-visit of the same strategy in a spiral way, the teachers become more competent and learn generalise it to new topics.
- *Coherence Principle.* Through training and personal experiences, teachers acquire many practical tips and strategies (Posamentier, Hartman, & Kaiser, 1998). Effective teaching results from integrating these strategies into coherent teaching acts. Teacher educators can assist this integration by analysing the research base of individual strategies, ensuring that the teachers take time and effort to acquire them, and finally scaffolding how to combine them into coordinated teaching styles.
- *Activity Principle.* The training sessions should include hands-on activities that engage the teachers in experimenting with the strategies with other participants.
- *Local Relevance and Global Perspective Principle.* Many effective instructional techniques can be found in East and West education systems. These are now readily shared across countries through international comparative studies such as TEDS-M, TIMSS, and PISA, and dedicated websites of professional bodies.

Teachers are encouraged to combine global techniques with local materials and praxis to enrich their lessons. The two resource books edited by Lee and Lee (2009a, 2009b) show how mathematics educators in Singapore have used this approach to design activities for pre-service teacher education.

- *Guidance and Constructivist Principle.* It is impossible for teacher educators to teach everything that teachers need to master. As teachers are adult learners, it is feasible to guide them to learn new techniques on their own either individually or with peers. Under the constructivist principle, teachers can be guided to work out the details based on theories, test the new skill in lessons, monitor critical events during trials, reflect on the outcomes, share experience with colleagues, and obtain feedback from them both formally and informally. This experimentation and sharing will evoke mutual support among participating teachers under various contexts, such as workshops and in-service courses. When this happens across schools, a networked learning community is established.

3.3 *Research own practices*

Teachers can research into their own practices in three different ways: enrol in award-bearing courses on action research, conduct action research, and join a research project (Wong, 2013a).

When teachers enrol in action research courses, they have to complete a project as part of course assessment. Many good books are available for use in such courses, and this huge literature attests the wide acceptance in many countries of action research as a powerful form of teacher professional development. In general, it is easier to learn research designs and standard data analyses but harder to acquire new teaching strategies to be trialled in the action research project. Hence, these courses must cover both research methods and subject-specific pedagogy in order that teachers can derive the purported benefits.

Having attended action research courses, teachers should continue to conduct action research individually or collaboratively in school-based teams. Collaborative action research can develop into communities of practice, learning circles, or professional learning communities (Chua, 2009). The following observations are based on action research reports published in Singapore. They highlight several shortcomings which must be addressed in order to use action research to bring about teacher growth in pedagogy.

- There is a strong preference to report mostly quantitative findings, highlighting statistical significance and effect sizes about student test scores. These results provide a fair measure of the *success* of the projects, but the findings cannot be generalised because the samples are usually very small and are not representative.
- Action research is basically about teachers trying to understand and resolve localised teaching or learning issues. This can be better achieved by analysing qualitative data rather than quantitative ones. Detailed analysis of student mathematical solutions is a common and potent way to understand their thinking in terms of misconceptions or creative solutions, but this kind of results are seldom published in these reports.
- Most teachers do not collect data about their in-class practices, and as a result, it is not possible to determine whether or not the projects had been implemented according to the plan or how a lack of implementation fidelity may affect the findings. As an illustration, suppose a teacher is interested in studying the use of non-examples in his or her lessons. These non-examples may have been included in the lesson plans, but this does not mean that the teacher will actually use some or all of them during the lessons. There are also different ways of using non-examples: explain them, use them to challenge students' claims, ask students to make up their own non-examples, and the like. Data collected about which non-examples are used, when, and for how long, are essential ingredients for subsequent reflection about the effects of this strategy. Without such data, the teachers cannot gain in-depth knowledge about their in-class practices. Hence, growth in pedagogical content knowledge is limited.
- Many reports include only positive results without any indication of whether negative findings were found or how implementation

problems had been overcome. This gloss does not give a balanced description of the real situations and is not consistent with the mindset of learning from mistakes and flaws. Sarason (1993) wrote that "Teachers not only had to learn a new way of thinking, but they had to unlearn old ways" (p. 106), and this unlearning can be initiated only when teachers realise their shortcomings and intend to change them.

- It seems that many projects were not replicated after the first attempt, which usually has teething problems. Thus, the teacher researchers do not avail themselves of the opportunity to learn from further attempts to hone their teaching skills, which is the main goal of participation in action research projects. Indeed, it takes considerable time and practice for teachers to learn a new teaching technique and to be able to integrate it into their teaching routines. Repeated attempts in different action research cycles are required to achieve this mastery.

Finally, the third way for teachers to learn from research is to become active members of research teams led by academic researchers. In this case, they can learn much about education theories, methodology, analyses, and report writing. In addition to acquiring these skills and knowledge, they will appreciate the challenges to link theory, practice, and research to achieve desirable curriculum goals.

4 Framework of Teacher Professional Development

It is not surprising that the main focus of teacher development is to improve teaching which is expected to lead to improved learning. However, research in teacher education has not provided definitive answers to this much sought-after link. Figure 10.2 shows several mediating factors between teacher professional development and student learning outcomes. It is modified from the frameworks by Clarke and Hollingsworth (2002), Coleman (2009), and Guskey (2002).

Teachers' participation in professional development is postulated to lead to changes in their mathematics content knowledge (MCK), maths-based pedagogy (MPCK), generic pedagogy (GP), and beliefs about teaching and learning. These changes are then reflected in changes in their classroom practices, which finally impact on student learning. For example, Wenglinsky (2002) reported that teachers who had learned to use concrete materials in mathematics lessons tended to produce stronger student achievement.

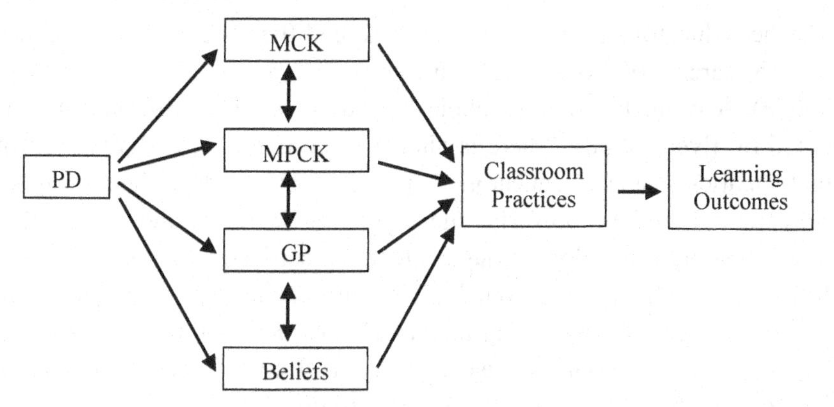

Figure 10.2. Mediating factors between teacher's professional development (PD) and student learning outcomes. MCK = Maths content knowledge; MPCK = Maths pedagogical content knowledge (maths-based pedagogy); GP = Generic pedagogy

The links in the framework differ in their influences (not highlighted in the diagram), and these influences may change over time. For example, Darling-Hammond (2000) suggested that teacher content knowledge may be important for initial years of teaching but its impact may be less than pedagogical knowledge in subsequent years. Furthermore, reciprocal influences exist among the four mediating factors (MCK, MPCK, GP, and beliefs).

Although the framework suggests that changes in beliefs lead to changes in classroom practices, some educators think otherwise: teachers first change their classroom practices, perhaps required by workshop facilitators, and this leads to changes in their beliefs when they observe improvement in student achievement. These alternative views highlight the complexity of the effects of professional development. Few research studies are able to unpack this complexity. It is not surprising that a recent US review concluded that "the professional development that takes place does not have an effect on student learning" (DeMonte, 2013, p.4). One way forward is to focus on the expertise and practice of teacher educators.

5 Professional Development of Teacher Educators

Teacher educators are teachers' teachers, and they "occupy the pinnacle of a hierarchy of power and authority" in teacher education (Wong, 2001b). Not much is known about how they develop professionally, as noted by Even (2008): "there is almost no research on the education of mathematics teacher educators" (p. 57). As academicians, teacher educators should be powerful metacognitive learners, and it is highly likely that they develop through attending conferences, consulting the literature, conducting research individually and in teams, working with school teachers on fine-tuning the latter's instructional practices, writing papers, teaching courses (teaching as a form of learning), having informal exchanges with colleagues, and so on.

The TEDS-M study (Tatto et al., 2012) reported findings about "whether teacher educators are being appropriately prepared" (p. 201). About 5200 teacher educators from 17 countries completed a survey about their qualifications. Percentages of teacher educators holding doctoral degrees in mathematics ranged from 7% to 60% across the countries; the percentages with doctoral degrees in mathematics pedagogy were similar, from 7% to 40%; those with school teaching experiences from 20% to 90%. These wide variations in qualifications may account for the different types and quality of teacher education programmes offered in these countries. These teacher educators tended "to endorse the pattern of beliefs described as conceptual or cognitive-constructionist in orientation" (p. 204) such as "mathematics is a process of inquiry and that it is best learned by active student involvement" (p. 203). Teacher educators from Singapore reported that they frequently encouraged pre-service teachers to ask questions during class time and to participate in whole-class discussions. But the pre-service teachers reported that these interactive learning experiences occurred much less frequently. This suggests a mismatch in perceptions between pre-service teachers and their teacher educators on the nature of teacher education programmes. At this stage, it is not possible to unpack the effects of this mismatch, but teacher educators are in the best position to change their practices to narrow this perceptual difference. Further research is called for.

6 Concluding Remarks

Providing professional development to teachers throughout their career is expensive in terms of funding, time, and resources. But this must be done because teachers have to acquire new content knowledge, teaching skills, and professional beliefs to cope with the ever changing education demands brought about by new local policies, global technologies, theories of psychology and neuroscience, and student characteristics. In this chapter, we have considered the scope of teacher professional activities and important principles used to organise these activities. Current research has not provided definitive findings about the effects of these activities on student learning outcomes, and it is hypothesised that these effects are mediated by several factors. Despite these uncertainties, teachers should strive to become metacognitive in finding their own pathway to become *better* teachers as they continue to enhance their toolkit in teaching and share it with colleagues, along the graduate-proficient-highly accomplished-lead continuum. Stakeholders in many countries are "still learning how to ensure that professional development delivers the results we desire" (DeMonte, 2013, p. 19). They can draw lessons from how behavioural economics have been successfully applied in non-education sectors to effectively influence people's decision making through small, concrete, and implementable changes under day-to-day circumstances, as documented in many examples in the recent World Bank report (2015). Thus, metacognitive teachers should strive to make small changes to their teaching practices, teacher educators can assist by researching the effects of these small changes, and policy-makers ought to provide the necessary support and rewards to spur this effort. Ultimately, the collaboration among different stakeholders must lead to effective teaching to advance student mathematics learning for the benefits of the individual students and their society.

7 References

The list below consists of only items that I have cited in the main text. It is obvious that much significant work in mathematics education made by important scholars and educators in the field has not been included here. I am confident that the interested readers will be able to find the missing pieces and use them to build their own schema about the three types of mathematics curriculum discussed in this book and to use this knowledge to strengthen their praxis. Enjoy your pursuit!

Advisory Committee on Mathematics Education (ACME, 2002). *Continuing professional development for teachers of mathematics.* Retrieved from http://www.acme-uk.org/media/1463/continuing%20professional%20development%20for%20teachers%20of%20mathematics.pdf

Afamasaga-Fuata'I, K. (Ed.). (2009). *Concept mapping in mathematics: Research into practice*. New York, NY: Springer.

Aiken, L.R., Jr. (1970). Attitudes toward mathematics. *Review of Educational Research, 40*, 551-596.

Akinwunmi, K., Höveler, K., & Schnell, S. (2014). On the importance of subject matter in mathematics education: A conversation with Erich Christian Wittmann. *Eurasia Journal of Mathematics, Science & Technology Education*, 10(4), 357-363.

Alfieri, L., Brooks, P.J., Aldrich, N.J., & Tenenbaum, H.R. (2011). Does discovery-based instruction enhance learning? *Journal of Educational Psychology, 103*(1), 1-18. Retrieved from http://www.apa.org/pubs/journals/features/edu-103-1-1.pdf

American Educational Research Association (AERA) (2005). *Teaching teachers: Professional development to improve student achievement: Essential information for education policy, 3*(1). Washington, DC: Author.

American Educational Research Association (AERA), American Psychological Association (APA), & the National Council on Measurement in Education (NCME). (1999). *Standards for educational and psychological testing.* Washington, DC: AERA.

Anderson, J.R., Reder, L.M., & Simon, H.A. (1999). *Applications and misapplications of cognitive psychology to mathematics education.* Retrieved from ERIC database. (ED439007)

Anderson, L.W., & Krathwohl, D.R. (Eds.). (2001). *A taxonomy for learning, teaching and assessment: A revision of Bloom's taxonomy of educational objectives* (complete ed.). New York, NY: Longman.

Ang, K.C. (2009). *Mathematical modelling in the secondary and junior college classroom*. Singapore: Prentice Hall.

Arem, C. (2010). *Conquering math anxiety: A self-help workbook* (3rd ed.). Pacific Grove, CA: Brooks/Cole.

Arrowsmith-Young, B. (2012). *The woman who changed her brain: How I left my learning disability behind and other stories of cognitive transformation*. New York, NY: Simon & Schuster Paperbacks.

Artmann, B. (1999). *Euclid: The creation of mathematics*. New York, NY: Springer.

Ashlock, R.B. (2010). *Error patterns in computation: Using error patterns to help each student learn* (10th ed.). Boston, MA: Allyn & Bacon.

Askew, M., Brown, M., Rhodes, V., Johnson, D., & William, D. (1997). *Effective teachers of numeracy: Final report. Report of a study carried out for the Teacher Training Agency*. London: King's College, University of London.

Association for Supervision and Curriculum Development (ASCD). (2014). *A lexicon of learning: What educators mean when they say...* Available from www.ascd.org/Publications/Lexicon-of-Learning/table-of-contents.aspx

Australian Education Council (1990). *A national statement on mathematics for Australian schools*. Carlton, Victoria: Curriculum Corporation.

Ausubel, D.P. (1968). *Educational psychology: A cognitive view*. New York, NY: Holt, Rinehart and Winston.

Ayres, I. (2007). *Super crunchers: Why thinking by numbers is the new way to be smart*. New York, NY: Bantam Books.

Baird, J. R., & Northfield, J. R. (1992). *Learning from the PEEL experience*. Melbourne: Monash University.

Bakken, J.P. (Ed.). (2012). *Response to Intervention in the core content areas: A practical approach for educators*. Waco, TX: Prufrock Press.

Ball, D., Lubienski, S.T., & Mewborn, D.S. (2001). Research on teaching mathematics: The unsolved problem of teachers' mathematical knowledge. In V. Richardson, (Ed.), *Handbook of research on teaching* (pp. 433-456). Washington, DC: American Educational Research Association.

Ball, D., Thames, M.H., & Phelps, G. (2008). Content knowledge for teaching: What makes it special? *Journal of Teacher Education, 59*(5), 389-407.

Barnard, T. (2005). *Hurdles and strategies in the teaching of algebra*. London: Mathematical Association.

Bell, A. (1993). Some experiments in diagnostic teaching. *Educational Studies in Mathematics, 24*(2), 115-137.

Bell, A. (1997). Purpose and awareness. In M. Hejný & J. Novotná (Eds.), *Proceedings of the European Research Conference on Mathematical Education* (p. 5–11). Prague: Charles University.

Benjamin, H.R.W. (2004). *The saber-tooth curriculum.* New York, NY: McGraw-Hill. (Original work published 1939)

Bennett, N. (1976). *Teaching styles and pupil progress.* London: Open Books.

Berry, J. & Houston, K. (1995). *Mathematical modelling.* London: Edward Arnold.

Beswick, K. (2006). The importance of teachers' beliefs. *Australian Mathematics Teacher, 62*(4), 17-22.

Biggs, J.B. (1987). *Student approaches to learning and studying.* Hawthorn: Australian Council for Educational Research.

Biggs, J.B. (1994). What are effective schools? Lessons from East and West. *Australian Educational Researcher, 21, 19-39.*

Birken, M. (1986). Teaching students how to study mathematics: A classroom approach. *Mathematics Teacher, 79*(6), 410–413.

Bishop, A.J. (1977). Is a picture worth a thousand words? *Mathematics Teaching, 81,* 32-35.

Bishop, A.J. (1991). *Mathematical enculturation: A cultural perspective on mathematics education.* Norwell, MA: Kluwer Academic Publishers.

Bloom, B.S. (Ed.). (1956). *Taxonomy of educational objectives: The classification of educational goals.* New York, NY: Longmans, Green.

Borasi, R. (1994). Capitalising on errors as "springboards for inquiry". *Journal for Research in Mathematics Education, 25*(2), 166–208.

Borman, K.M., Kersaint, G., Cotner, B., Lee, R., Boydston, T., Uekawa, K., Kromrey, W., Baber, M.Y., & Barber, J. (2005). *Meaningful urban education reform: Confronting the learning crisis in mathematics and science*. New York, NY: State University of New York Press.

Brenner, M.E., & Moschkovich, J.N. (Eds.). (2002). *Everyday and academic mathematics in the classroom.* Reston, VA: National Council of Teachers of Mathematics.

Brownell, W.A. (2004). The place of meaning in the teaching of arithmetic. In T.P. Carpenter, J.A. Dossey & J.L. Koehler (Eds.), *Classics in mathematics education research* (pp. 9-14). Reston, VA: National Council of Teachers of Mathematics. (Original work published 1947)

Bruner, J.S. (1960). *The process of education.* Cambridge, MA: Harvard University Press.

Bruner, J.S. (1964a). The course of cognitive growth. *American Psychologist, 19*, 1–15.

Bruner, J.S. (1964b). Some theorems on instruction illustrated with reference to mathematics. In E.R. Hilgard (Ed.), *Theories of Learning and Instruction: NSSE Yearbook* (pp. 306-335). Chicago, IL: National Society for the Study of Education.

Buoncristiani, A.M., & Buoncristiani, P. (2012). *Developing mindful students, skillful thinkers, thoughtful schools*. Thousand Oaks, CA: Corwin.

Burghes, D., Galbraith, P., Price, N., & Sherlock, A. (1996). *Mathematical modelling.* London: Prentice Hall.

Butterworth, B. (1999). *What counts: How every brain is hardwired for math.* New York, NY: Free Press.

Callahan, L.G., & Garofalo, J. (1987). Metacognition and school mathematics. *Arithmetic Teacher, 34*(9), 22-23.

Chan, C.M.E. (2012). *Techniques in solving higher-order thinking word problems: Upper primary, teacher's notes.* Singapore: Star Publishing.

Chapin, S.H., O'Connor, C., & Anderson, N.C. (2009). *Classroom discussions: Using math talk to help students learn, Grades 1-6* (2nd ed.). Sausalito, CA: Math Solutions Publications.

Charles, R., Lester, F., & O'Daffer, P. (1987). *How to evaluate progress in problem solving.* Reston, VA: National Council of Teachers of Mathematics.

Chipperfield, B. (2004). *Cognitive load theory and instructional design.* Retrieved from http://www.usask.ca/education/coursework/802papers/chipperfield/index.htm

Chua, B.L., & Hoyles, C. (2014). Generalisation of linear figural patterns in secondary school mathematics. *The Mathematics Educator, 15*(2), 1-30.

Chua, P.H. (2009). Learning communities: Roles of Teachers Network and zone activities. In K.Y. Wong, P.Y. Lee, B. Kaur, P.Y. Foong, & S.F. Ng (Eds.), Mathematics education: The Singapore journey (pp. 85-103). Singapore: World Scientific.

Clark, R.E. (2006). Not knowing what we don't know: Reframing the importance of automated knowledge for educational research. In G. Clarebout & J. Elen (Eds.), *Avoiding simplicity, confronting complexity: Advances in studying and designing (computer-based) powerful learning environments* (pp. 3-14). Rotterdam: Sense Publishers.

Clarke, D. (2007). Ten key principles from research for the professional development of mathematics teachers. In G.C. Leder & H.J. Forgasz (Eds.), *Stepping stones for the 21st century: Australasian mathematics education research* (pp. 27-39). Rotterdam: Sense Publishers. (Original work published 1994)

Clarke, D., & Hollingsworth, H. (2002). Elaborating a model of teacher professional growth. *Teaching and Teacher Education, 18*(8), 947–967.

Clements, D.H., & Battista, M.T. (with Sarama, J.). (2001). *Logo and geometry.* Reston, VA: National Council of Teachers of Mathematics.

Clements, D.H., & McMillen, S. (1999). Rethinking "concrete" manipulatives. In A.R. Teppo (Ed.), *Reflecting on practice in elementary school mathematics: Readings from NCTM's school-based journals and other publications* (pp. 140-149). Reston, VA: National Council of Teachers of Mathematics.

Cockcroft, W.H. (1982). *Mathematics counts.* London: HMSO.

Coe, R., Aloisi, C., Higgins, S., & Major, L.E. (2014). *What makes great teaching? Review of the underpinning research.* London: The Sutton Trust. Retrieved from http://www.suttontrust.com/wp-content/uploads/2014/10/What-makes-great-teaching-FINAL-4.11.14.pdf

Colburn, A. (2003). *The lingo of learning: 88 education terms every science teacher should know.* Arlington, VA: NSTA Press.

Coleman, D.C. (2009). Building capacity for improving teaching quality. In J.S. Kettlewell & R.J. Henry (Eds.), *Increasing the competitive edge in math and science* (pp. 85-100). Lanham, MD: Rowman & Littlefield Education.

Colman, A.M. (2006). *Oxford dictionary of psychology.* Oxford: Oxford University Press.

Common Core State Standards Initiative (CCSSI). (2010). *Common Core State Standards for mathematics.* Washington, DC: National Governors Association Center for Best Practices and Council of Chief State School Officers (CCSSO). Retrieved from http://www.corestandards.org/the-standards

Converse, J.M., & Presser, S. (1986). *Survey questions: Handcrafting the standardized questionnaire.* Newbury Park, CA: Thousand Oaks, CA: Sage Publishers.

Cooke, H. (2003). *Success with mathematics.* London: Routledge.

Cooney, T.J. (2001). Considering the paradoxes, perils, and purposes of conceptualizing teacher development. In F.L. Lin & T.J. Cooney (Eds.), *Making sense of mathematics teacher education* (pp. 9-31). Dordrecht: Kluwer Academic Publishers.

Cooper, H. (2008). *Research brief: Homework, what the research says?* Reston, VA: National Council of Teachers of Mathematics.

Corkill, P. (1996). Enhancing effective student learning in mathematics: A collaborative approach. In H. Forgasz, T. Jones, G. Leder, J. Lynch, K. Maguire & C. Pearn (Eds.), *Mathematics: Making connections* (pp. 31–41). Melbourne: Mathematical Association of Victoria.

Courant, R., & Robbins, H. (1947). *What is mathematics? An elementary approach to ideas and methods* (4th ed.). London: Oxford University Press.

Crawford, K., Gordon, S., Nicholas, J., & Prosser, M. (1998). Qualitatively different experiences of learning mathematics at university. *Learning and Instruction, 8*(5), 455–468.

Cross, M., & Moscardini, A.O. (1985). *Learning the art of mathematical modelling.* Chichester: Ellis Horwood.

Csikszentmihalyi, M. (1997). *Finding flow: The psychology of engagement with everyday life.* New York, NY: Basic Books.

Cuoco, A. (2001). Mathematics for teaching. *Notices of the American Mathematical Society, 48*(2), 168-174.

D'Ambrosio, U. (2006). *Ethnomathematics link between traditions and modernity.* Rotterdam, the Netherlands: Sense Publishers.

Darling-Hammond, L. (2000). Teacher quality and student achievement: A review of state policy evidence. *Education Policy Analysis Archives, 8*(1). Retrieved from http://epaa.asu.edu/ojs/article/view/392/515

Davis, R.B., Maher, C.A., & Noddings, N. (Eds.). (1990). *Constructivist views on the teaching and learning of mathematics*. Reston, VA: National Council of Teachers of Mathematics.

de Bono, E. (1992). *Teach your child how to think.* London: Penguin.

DeMonte, J. (2013). *High-quality professional development for teachers: Supporting teacher training to improve student learning.* Washington, DC: Center for American Progress.

Department for Children, Schools and Families (DCSF). (2010). *Assessing pupils' progress: A teachers' handbook.* London: Author. Retrieved from http://webarchive.nationalarchives.gov.uk/20110202093118/http://nationalstrategies.standards.dcsf.gov.uk/node/259613

Desoete, A., & Veenman, M. (Eds.). (2006). *Metacognition in mathematics education.* New York, NY: Nova Science Publishers.

Devlin, K.J. (1997). *Mathematics, the science of patterns: The search for order in life, mind, and the universe.* New York, NY: Scientific American Library.

Dewdney, A.K. (1993). *200% of nothing: An eye-opening tour through the twists and turns of math abuse and innumeracy.* New York, NY: John Wiley & Sons.

Di Teodoro, S., Donders, S., Kemp-Davidson, J., Robertson, P., & Schuyler, L. (2011). Asking good questions: Promoting greater understanding of mathematics through purposeful teacher and student questioning. *Canadian Journal of Action Research, 12*(2), 18-29.

Dienes, Z.P. (1964). *Building up mathematics* (2nd ed.). London: Hutchinson Educational.

Dindyal, J. (2009). *Applications and modelling for the primary mathematics classroom.* Singapore: Prentice Hall.

Dindyal, J., Tay, E.G., Toh, T.L., Leong, Y.H., & Quek, K.S. (2012). Mathematical problem solving for everyone: A new beginning. *The Mathematics Educator, 13*(2), 1-20.

Doidge, N. (2007). *The brain that changes itself: Stories of personal triumph from the frontiers of brain science.* New York, NY: Penguin Books.

Donnelly, K., & Wiltshire, K. (2014). *Review of the Australian curriculum: Final report.* Canberra: Australian Government Department of Education. Retrieved from https://docs.education.gov.au/system/files/doc/other/review_of_the_national_curriculum_final_report.pdf

Dweck, C.S. (2006). *Mindset: The new psychology of success.* New York, NY: Random House.

Dynarski, M., Agodini, R., Heaviside, S., Novak, T., Carey, N., Campuzano, L., Means, B., Murphy, R., Penuel, W., Javitz, H., Emery, D., & Sussex, W. (2007). *Effectiveness of reading and mathematics software products: Findings from the First Student Cohort. Report to Congress.* Washington, DC: U.S. Department of Education, Institute of Education Sciences. Retrieved from http://ies.ed.gov/ncee/pdf/20074005.pdf

Earl, L.M. (2013). *Assessment as learning: Using classroom assessment to maximize student learning* (2nd ed.). Thousand Oaks, CA: Corwin Press.

Ee, H.W. (2014). *Experiences of learning Engineering Mathematics under different contexts in a Singapore polytechnic.* Unpublished doctoral thesis, Nanyang Technological University, Singapore.

Elder, L., & Paul, R. (2005). *The miniature guide to the art of asking essential questions.* Dillon Beach, CA: The Foundation for Critical Thinking.

Ellenberg, J. (2014). *How not to be wrong: The power of mathematical thinking.* New York, NY: Penguin Press.

Entrekin, V. S. (1992). Mathematical mind mapping. *The Mathematics Teacher, 85*(6), 444–445.

Enyedy, N. (2014). *Personalized Instruction: New interest, old rhetoric, limited results, and the need for a new direction for computer-mediated learning.* Boulder, CO: National Education Policy Center. Retrieved from http://nepc.colorado.edu/publication/personalized-instruction

Ernest, P. (1989). What's the use of LOGO? In P. Ernest (Ed.), *Mathematics teaching: The state of the art* (pp. 33-44). London: Falmer Press.

Ernest, P. (1991). *The philosophy of mathematics education.* London: Falmer Press.

Ernest, P. (1998). Recent developments in mathematical thinking. In R. Burden & M. Williams (Eds.), *Thinking through the curriculum* (pp. 113-134). London: Routledge.

Esquith, R. (2013). *Real talk real teachers: Advice for teachers from rookies to veterans: "No retreat, no surrender!".* New York, NY: Viking.

Even, R. (2008). Facing the challenge of educating educators to work with practising mathematics teachers. In B. Jaworski & T. Wood (Eds.), *The mathematics teacher educator as a developing professional* (pp. 57-73). Rotterdam: Sense Publishers.

Everitt, B.S., Landau, S., & Leese, M. (2001). *Cluster analysis* (4th ed.). Oxford: Oxford University Press.

Fan, L.H., & Cheong, N.P. (2003). *The sources of teachers' pedagogical knowledge: The case of Singapore* (Unpublished Technical Report). Singapore: Nanyang Technological University, National Institute of Education.

Fan, L., Quek, K.S., Koay, P.L., Ng, C.H.J.D., Pereira-Mendoza, L., Yeo, S.M., Tan-Foo, K.F., Teo, S.W., & Zhu, Y. (2008). *Integrating new assessment strategies into mathematics classrooms: An exploratory study in Singapore primary and secondary schools* (Final Research Report). Singapore: National Institute of Education, Centre for Research in Pedagogy and Practice. Retrieved from http://eprints.soton.ac.uk/174457/1/CRP24_03FLH_FinalResRpt.pdf

Felmer, P., Lewin, L., Martínez, S., Reyes, C., Varas, L., Chandía, E., Dartnell, P., López, A., Martínez, C., Mena, A., Ortíz, A., Schwarze, G., & Zanocco, P. (Eds.). (2014). *Primary mathematics standards for pre-service teachers in Chile: A resource book for teachers and educators.* Singapore: World Scientific.

Fennema, E. & Franke, M.L. (1992). Teachers' knowledge and its impact. In D.A. Grouws (Ed.), *Handbook of Research on Mathematics Teaching and Learning* (pp. 147-164). New York: Macmillan & National Council of Teachers of Mathematics.

Ferrucci, B.J., Yeap, B.H., & Carter, J.A. (2003). A modeling approach for enhancing problem solving in the middle grades. *Mathematics Teaching in the Middle Schools, 8*(9), 470-475.

Flavell, J.H. (1976). Metacognitive aspects of problem solving. In L.B. Resnick (Ed.), *The nature of intelligence* (pp. 231-235). Hillsdale, NJ: Lawrence Erlbaum.

Fletcher, T.J., Milner, W.W., & Watson, F.R. (1990). *Logo and mathematics.* Staffordshire: University of Keele.

Flores, M.M. (2010). Using the concrete-representational-abstract sequence to teach subtraction with regrouping to students at risk for failure. *Remediation and Special Education, 31*(3), 195-207.

Fogarty, R. (1994). *How to teach for metacognitive reflection.* Palantine, IL: IRI/Skylight.

Fong, H.K. (1993). *Challenging mathematical problems for primary schools: The model approach.* Sydney: Kingsford Educational Services.

Fong, H.K. (1998). *Solving challenging mathematical problems for primary schools: The heuristic approach.* Sydney: Kingsford Educational Services.

Fong, W.C. (1986). *The use of Specific Instructional Objectives for effective teaching and testing.* Singapore: Ministry of Education.

Foong, P.Y. (2009a). Problem solving in mathematics. P. Y. Lee & N. H. Lee (Eds.), *Teaching primary school mathematics: A resource book* (2nd. ed.) (pp. 54-81). Singapore: McGraw-Hill Education (Asia).

Foong, P.Y. (2009b). Review of research on mathematical problem solving in Singapore. In K.Y. Wong, P.Y. Lee, B. Kaur, P.Y. Foong, & S.F. Ng (Eds.), *Mathematics education: The Singapore journey* (pp. 263-300). Singapore: World Scientific.

Forsten, C. (2010). *Step-by-step model drawing: Solving word problems the Singapore way.* Peterborough, NH: Crystal Springs Books.

Fraenkel, J.R., & Wallen, N.E. (1993). *How to design and evaluate research in education* (3rd ed.). New York, NY: McGraw-Hill.

Frenzel, A.C., Goetz, T., Lüdtke, O., Pekrun, R., & Sutton, R.E. (2009). Emotional transmission in the classroom: Exploring the relationship between teacher and student enjoyment. *Journal of Educational Psychology, 101*(3), 705-716.

Gagné, R.N. & Briggs, L.J., & Wager, W.W. (1992). *Principles of instructional design* (4th ed.). Belmont, CA: Wadsworth.

Galbraith, P. (1982). The mathematical vitality of secondary mathematics graduates and prospective teachers: A comparative study. *Educational Studies in Mathematics, 13*, 89 – 112.

Galbraith, P. (1988). Mathematics education and the future: A long wave view of change. *For the Learning of Mathematics, 8*(3), 27–33.

Gardiner, T. (1987). Mathematical methods: Does it exist? *Mathematical Gazette, 71*, 265-271.

Gardner, H. (1991). *The unschooled mind.* London: Fontana Press.

Gardner, H. (2004). *Changing minds: The art and science of changing our own and other people's minds.* Boston, MA: Harvard Business School Press.

Gates, P., & Noyes, A. (2014). School mathematics as social classification. In D. Leslie & H. Mendick (Eds.), *Debates in mathematics education* (pp. 38-48). London: Routledge.

Gibboney, R.A. (with Webb, C.W.). (1998). *What every great teacher knows: Practical principles for effective teaching.* Brandon, VT: Holistic Education Press.

Gigerenzer, G. (2002). *Reckoning with risk: Learning to live with uncertainty.* London: Penguin Books.

Glass, G.V. (1976). Primary, secondary, and meta-analysis of research. *Educational Researcher, 5*, 3-8.

Gningue, S. M. (2006). Students working within and between Representations: An application of Dienes's Variability Principles. *For the Learning of Mathematics, 26*(2), 41-47.

Goh, S.P. (2009). *Primary 5 pupils' difficulties in using the model method for solving complex relational word problems.* Unpublished Masters dissertation, Nanyang Technological University, Singapore.

Gojak, L. (2011). *What's your math problem!?!: Getting to the heart of teaching problem solving.* Huntington Beach, CA: Shell Education.

Goldin-Meadow, S. (2003). *Hearing gesture: How our hands help us think.* Cambridge, MA: Harvard University Press.

Gray, E.M., & Tall, D.O. (1994). Duality, ambiguity, and flexibility: A proceptual view of simple arithmetic. *Journal for Research in Mathematics Education, 25*(2), 116-141.

Gredler, M.E. (2005). *Learning and instruction: Theory into practice* (5th ed.). Upper Saddle River, NJ: Pearson/Merrill Prentice Hall.

Grootenboer, P., Lomas, G., & Ingram, N. (2008). The affective domain and mathematics education. In H. Forgasz, A. Barkatsas, A. Bishop, B. Clarke, S. Keast, W.T. Seah & P. Sullivan (Eds.), *Research in mathematics education in Australasia 2004-2007* (pp. 255-269). Rotterdam: Sense Publishers.

Guskey, T.R. (2002). Professional development and teacher change. *Teachers and Teaching: Theory and practice, 8*(3/4), 381-391.

Hansen, A. (2011). *Children's errors in mathematics: Using common misconceptions in primary schools* (2nd ed.). Exeter: Learning Matters Ltd.

Hansen, A., & Vaukins, D. (2011). *Primary mathematics across the curriculum.* Exeter: Learning Matters.

Hargreaves, A., Earl, L., Moore, S., & Manning, S. (2001). *Learning to change: Teaching beyond subjects and standards.* San Francisco, CA: Jossey-Bass.

Hart K. M. (Ed.). (1981). *Children's understanding of mathematics 11-16.* London: John Murray.

Harvey B. (1987). *Computer science Logo style. Vol. 3.* Cambridge, MA: MIT Press.

Hattie, J. (2005, August). *What is the nature of evidence that makes a difference to learning?* Paper presented at the Australian Council for Educational Research Conference "Using Data to Support Learning," Melbourne, Australia. Retrieved from http://www.acer.edu.au/workshops/documents/Hattie.pdf

Hattie, J.A.C. (2009). *Visible learning: A synthesis of over 800 meta-analyses relating to achievement.* London: Routledge.

Hattie, J.A.C. (2012). *Visible learning: Maximizing impact on learning.* London: Routledge.

Hawkridge, D., Jaworski, J., & McMahon, H. (1990). *Computers in third-world schools: Examples, experience and issues.* London: Macmillan.

Haylock, D.W. (1984). A mathematical think board. *Mathematics Teaching, 108*, 4-5.

Healy, J.M. (1999). *Failure to connect: How computers affect our children's mind and what we can do about it.* New York, NY: Simon & Schuster.

Heeffer, A. (na). *Negative numbers as an epistemic difficult concept: Some lessons from history.* Retrieved from logica.ugent.be/albrecht/thesis/HPM2008.pdf

Henning, J.E. (2008). *The art of discussion-based teaching: Opening up conversation in the classroom.* New York, NY: Routledge.

Hiebert, J., & Lefevre, P. (1986). Conceptual and procedural knowledge in mathematics: An introductory analysis. In J. Hiebert (Ed.), *Conceptual and procedural knowledge: The case of mathematics* (pp. 1-17). Hillsdale, NJ: Lawrence Erlbaum Associates.

Hill, H.C., Ball, D.L., & Schilling, S. (2008). Unpacking pedagogical content knowledge: Conceptualizing and measuring teachers' topic-specific knowledge of students. *Journal for Research in Mathematics Education, 39*(4), 372-400.

Hintz, A., & Kazemi, E. (2014). Talking about math. *Educational Leadership, 72*(3), 36-40.

Ho, C.S.Y. (1997). *A study of the effects of computer assisted instruction on the teaching and learning of transformation geometry.* Unpublished M Ed dissertation, National Institute of Education, Nanyang Technological University, Singapore.

Hogan, T.P. (2007). *Educational assessment: A practical introduction.* Hoboken, NJ: John Wiley & Sons.

Holton, D., & Clarke, D. (2006). Scaffolding and metacognition. *International Journal of Mathematical Education in Science & Technology, 37*(2), 127-143.

Houston, S. K., Blum, W., Huntley, I. D., & Neill, N.T. (Eds.). (1997). *Teaching and learning mathematical modelling: Innovation, investigation and applications.* Chichester: Albion Publishing.

Hoven, J., & Garelick, B. (2007). Singapore Math: Simple or complex? *Educational Leadership, 65*(3), 28-31.

Howson, A.G. (1983). *A review of research in mathematical education: Part C: Curricular development and curricular research. A historical and comparative view.* Windsor, Berks: NFER-Nelson.

Howson, A.G., Keitel, C., & Kilpatrick, J. (1981). *Curriculum development in mathematics.* Cambridge: Cambridge University Press.

Hoyles, C., & Lagrange, J. (Eds.). (2010). *Mathematics education and technology: Rethinking the terrain*. New York, NY: Springer.

Hoyles, C., & Noss, R. (Eds.). (1992). *Learning mathematics and Logo*. Cambridge, MA: MIT Press.

Huntley, I.D., & James, D.J.G. (Eds.). (1990). *Mathematical modelling: A source book of case studies*. Oxford: Oxford University Press.

Hyde, S. (1992). Negotiating mathematics. In G. Boomer, N. Lester, C. Onore, & J. Cook (Eds.), *Negotiating the curriculum: Educating for the 21st century* (pp. 53-66). London: Falmer Press.

Inglis, F., & Aers, L. (2008). *Key concepts in education.* London: Sage Publishers.

Jin, H. (2013). *Conceptual understanding of Grade 8 students about basic algebra and geometric shapes: Using concept map as an assessment technique*. Unpublished doctoral thesis, Nanyang Technological University, Singapore.

Jin, H.Y., & Wong, K.Y. (2011). Assessing conceptual understanding in mathematics with concept mapping. In B. Kaur & K.Y. Wong (Eds.), *Assessment in the mathematics classrooms: Association of Mathematics Educators Yearbook 2011* (pp. 67-90). Singapore: World Scientific.

Johansson, P., & Gärdenfors, P. (2005). Introduction to cognition, education, and communication technology. In P. Gärdenfors & P. Johansson (Eds.), *Cognition, education, and communication technology* (pp. 1-20). Mahwah, NJ: Lawrence Erlbaum Associates.

Kaplan, R., & Kaplan, E. (2007). *Out of the labyrinth: Setting mathematics free.* New York, NY: Oxford University Press.

Kaur, B. (2008). *Problem solving in the mathematics classroom (Secondary).* Singapore: National Institute of Education and Association of Mathematics Educators.

Kaur, B. (2011). Mathematics homework: A study of three grade eight classrooms in Singapore. *International Journal of Science and Mathematics Education, 9*(1), 187-206.

Kaur, B. (Ed.). (2013). *Nurturing reflective learners in mathematics: Association of Mathematics Educators Yearbook 2013*. Singapore: World Scientific.

Kaur, B., & Dindyal, J. (Eds.). (2010). *Mathematical applications and modelling: Yearbook 2010.* Singapore: World Scientific.

Kaur, B., & Yeap, B.H. (2009a). *Pathways to reasoning and communication in the primary school mathematics classroom: A resource for teachers by teachers.* Singapore: National Institute of Education.

Kaur, B., & Yeap, B.H. (2009b). *Pathways to reasoning and communication in the secondary school mathematics classroom: A resource for teachers by teachers.* Singapore: National Institute of Education.

Kho, T.H., Yeo, S.M., & Lim, J. (2009). *The Singapore model method for learning mathematics.* Singapore: Curriculum Planning & Development Division, Ministry of Education.

Kolata, G., & Hoffman, P. (Eds.). (2013). *The New York Times book of mathematics: More than 100 years of writing by the numbers.* New York, NY: Sterling.

Kohn, A. (2001). *Five reasons to stop saying "Good Job!"* Retrieved from http://www.alfiekohn.org/article/five-reasons-stop-saying-good-job/

Kounin, J.S. (1977). *Discipline and group management in classrooms.* Huntington, NY: R.E. Krieger.

Kramarski, B. & Mevarech, Z.R. (1997). Cognitive-metacognitive training within a problem-solving based Logo environment. *British Journal of Educational Psychology, 67*(4), 425-446.

Krpan, C.M. (2013). *Math expressions: Developing student thinking and problem solving through communication.* Toronto: Pearson Canada.

Lambdin, D.V., & Walcott, C. (2007). Changes through the years: Connections between psychological learning theories and the school mathematics curriculum. In W.G. Martin, M.E. Strutchens, & P.C. Elliott (Eds.), *The learning of mathematics: Sixty-ninth yearbook* (pp. 3-25). Reston, VA: National Council of Teachers of Mathematics.

Law, N. (2007). Comparing pedagogical innovations. In M. Bray, B. Adamson, & M. Mason (Eds.), *Comparative education research: Approaches and methods* (pp. 315-337). Hong Kong: University of Hong Kong.

Lee, J. (2009). Universals and specifics of math self-concept, math self-efficacy, and math anxiety across 41 PISA 2003 participating countries. *Learning and Individual Differences, 19,* 355-365.

Lee, K. & Ng, S.F. (2009). Solving algebra word problems: The roles of working memory and the model method. In K.Y. Wong, P.Y. Lee, B. Kaur, P.Y. Foong & S.F. Ng (Eds.), Mathematics education: The Singapore journey (pp. 204-226). Singapore: World Scientific.

Lee, N.H. (2009). Preparation of schemes of work and lesson plans. In P.Y. Lee & N.H. Lee (Eds.), *Teaching secondary school mathematics: A resource book (2nd. & updated ed.)* (pp. 337-356). Singapore: McGraw-Hill Education (Asia).

Lee, N.H., & Ng, K.E.D. (Eds.). (2015). *Mathematical modelling: From theory to practice.* Singapore: World Scientific.

Lee, P. Y. (2008). Sixty years of mathematics syllabus and textbooks in Singapore (1949-2005). In Z. Usiskin & E. Willmore (Eds.), *Mathematics curriculum in Pacific Rim countries – China, Japan, Korea, and Singapore* (pp. 85-94). Charlotte, NC: Information Age Publishing.

Lee, P.Y. & Lee, N.H. (Ed.). (2009a). *Teaching primary school mathematics: A resource book* (2nd ed.). Singapore: McGraw-Hill Education (Asia).

Lee, P.Y. & Lee, N.H. (Ed.). (2009b). *Teaching secondary school mathematics: A resource book* (2nd & updated ed.). Singapore: McGraw-Hill Education (Asia).

Leong, Y.H. (2003). Use of Geometer's Sketchpad in secondary schools. *The Mathematics Educator, 7*(2), 86-95.

Leong, Y.H. (2012). Presenting mathematics as connected in the secondary classroom. In B. Kaur & T.L. Toh (Eds.), *Reasoning, communication and connections in mathematics: Association of Mathematics Educators Yearbook 2012* (pp. 239-260). Singapore: World Scientific.

Leong, Y.H., Tay, E.G., Quek, K.S., Toh, T.L., Toh, P.C., Dindyal, J., Ho, F.H., & Yap, A.S.R. (2014). *Making Mathematics More Practical: Implementation in the Schools.* Singapore: World Scientific.

Leong, Y.H., Tay, E.G., Quek, K.S., Yap, S.F., Lee, H.T.C., Tong, C.L., Toh, W.Y.K., Xie, X.R., & Yeo, J.S.D. (2014). *MProVE negative numbers*. Singapore: National Institute of Education.

Leong. Y.H., Tay, E.G., Quek, K.S., Yap, S.F., Tong, C.L., Toh, W.Y.K., Xie, X.R., & Yeo, J.S.D. (2014). *MProVE number patterns*. Singapore: National Institute of Education.

Leong, Y.H., Yap, S.F., Teo, M.L.Y., Thilagam d/o Subramaniam, Irni Karen Bte Mohd Zaini, Quek, E.C., & Tan, K.L.K. (2010). Concretising factorisation of quadratic expressions. *Australian Mathematics Teacher, 66*(3), 19-24.

Lesh, R., Landau, M., & Hamilton, E. (1983). Conceptual models and applied mathematical problem-solving research. In R. Lesh & M. Landau (Eds.), *Acquisition of mathematics concepts and processes* (pp. 263-343). New York, NY: Academic Press.

Lesser, L. (2001). Representation of reversal: An exploration of Simpson's Paradox. In A. Cuoco & F. Curciol (Eds.), *The roles of representation in school mathematics* (pp.129-145). Reston, VA: National Council of Teachers of Mathematics.

Lew, H.C., & Jang, I.O. (2012). Logo project-based mathematics learning for communication, reasoning and connection. In B. Kaur & T.L. Toh (Eds.), *Reasoning, communication and connections in mathematics: Association of Mathematics Educators Yearbook 2012* (pp. 107-126). Singapore: World Scientific.

Lewis, A.B., & Mayer, R.E. (1987). Students' miscomprehension of relational statements in arithmetic word problems. *Journal of Educational Psychology, 79*(4), 363-371.

Li, Q., & Ma, X. (2010). A meta-analysis of the effects of computer technology on school students' mathematics learning. *Educational Psychology Review, 22*(3), 215-243.

Liebeck, P. (1984). *How children learn mathematics: A guide for parents and teachers.* London: Penguin.

Lim, S.K., & Wong, K.Y. (1989). Perspectives of an effective mathematics teacher. *Singapore Journal of Education*, Special Issue, 101-105.

Lim-Teo, S.K. (2009). Mathematics teacher education: Pre-service and in-service programmes. In K.Y. Wong, P.Y. Lee, B. Kaur, P.Y. Foong & S.F. Ng (Eds.), *Mathematics education: The Singapore journey* (pp. 48-84). Singapore: World Scientific.

Lim-Teo, S.K. (2012). The mathematical vitality of undergraduate mathematics student teachers in Singapore. *The Mathematics Educator, 14*(1&2), 1-20.

Lim-Teo, S.K., Chua, K.G., Cheang, W.K., & Yeo, J.K. (2007). The development of Diploma in Education student teachers' mathematics pedagogical content knowledge. *International Journal of Science and Mathematics Education, 5*(2), 237-261.

Long, C.T., DeTemple, D.W., & Millman, R.S. (2012). *Mathematical reasoning for elementary teachers* (6th ed.). Boston, MA: Addison Wesley.

Loveless, T. (2006). *How well are American students learning? With special sections on the achievement, the happiness factor in learning, and honesty in state test scores.* Washington, DC: Brookings Institute. Retrieved from http://www.brookings.edu/gs/brown/bc_report/2006/2006report.pdf

Ma, L. (1999). *Knowing and teaching elementary mathematics: Teachers' understanding of fundamental mathematics in China and the United States.* Mahwah, NJ: Lawrence Erlbaum Associates.

Maasβ, J., & O'Donoghue, J. (Eds.). (2011). *Real-world problems for secondary school mathematics students: Case studies.* Rotterdam, the Netherlands: Sense Publishers.

Mackenzie, D. (2012). *The universe in zero words: The story of mathematics as told through equations.* Princeton, NJ: Princeton University Press.

Manning, B.H., & Payne, B.D. (1996). *Self-talk for teachers and students: Metacognitive strategies for personal and classroom use.* Boston, MA: Allyn and Bacon.

Margenau, J., & Sentlowitz, M. (1977). *How to study mathematics.* Reston, VA: National Council of Teachers of Mathematics.

Martin, D., Paulsen, M., & Prata, S. (1985). *IBM PC and PCjr Logo programming primer.* Indianapolis, IN: H.W. Sams.

Mason, J.H. (1999). *Learning and doing mathematics* (2nd rev. ed.). York: QED.

Mason, J., & Johnston-Wilder, S. (Eds.). (2004). *Fundamental constructs in mathematics education.* London: Routledge-Falmer.

Mayer, R.E. (2006a). The role of domain knowledge in creative problem solving. In J.C. Kaufman & J. Baer (Eds.), *Creativity and reason in cognitive development* (pp. 145-158). Cambridge: Cambridge University Press.

Mayer, R.E. (2006b). Ten research-based principles of multimedia learning. In H.F. O'Neil & R.S. Perez (Eds.), *Web-based learning: Theory, research, and practice* (pp. 371-390). Mahwah, NJ: Lawrence Erlbaum Associates.

Mayer, R.E., & Moreno, R. (2003). Nine ways to reduce cognitive load in multimedia learning. *Educational Psychologist, 38*(1), 43-52. doi:10.1207/S15326985EP3801_6

McClure, J.R., Sonak, B., & Suen, H.K. (1999). Concept map assessment of classroom learning: Reliability, validity, and logistical practicality. *Journal of Research in Science Teaching, 36*(4), 475-492.

McNamara, D.S. (1995). Effects of prior knowledge on the generation advantage: Calculators versus calculation to learn simple multiplication. *Journal of Educational Psychology, 87*(2), 307-318.

Means, B., Toyama, Y., Murphy, R., Bakia, M., & Jones, K. (2010). *Evaluation of evidence-based practices in online learning: A meta-analysis and review of online learning studies.* Washington, DC: US Department of Education. Retrieved from www.ed.gov/about/offices/list/opepd/ppss/reports.html

Mevarech, Z., & Fridkin, S. (2006). The effects of IMPROVE on mathematical knowledge, mathematical reasoning and meta-cognition. *Metacognition and Learning, 1*(1), 85-97.

Ministry of Education, Singapore. (2000). *Mathematics syllabus (Primary).* Singapore: Author.

Ministry of Education, Singapore. (2012). *Primary mathematics: Teaching and learning syllabus.* Singapore: Author.

Mooney, C., Briggs, M., Fletcher, M., Hansen, A., & McCullouch, J. (2012). *Primary mathematics: Teaching theory and practice* (6th ed.). London: Learning Matters.

Morris, A.K., Hiebert, J., & Spitzer, S.M. (2009). Mathematical knowledge for teaching in planning and evaluating instruction: What can preservice teachers learn? *Journal for Research in Mathematics Education, 40*(5), 491 – 529.

Morrison, K. (2009). *Causation in educational research.* London: Routledge.

Mullis, I. V.S., Martin, M. O., & Foy, P. (2008). *TIMSS 2007 international mathematics report: Findings from IEA's Trends in International Mathematics and Science Study at the fourth and eighth grades.* Chestnut Hill, MA: TIMSS & PIRLS International Study Center, Lynch School of Education, Boston College.

Mullis, I. V.S., Martin, M. O., Foy, P., & Arora, A. (2012). *TIMSS 2011 International results in mathematics.* Chestnut Hill, MA & Amsterdam, the Netherlands: TIMSS & PIRLS International Study Center, Lynch School of Education, Boston College & International Association for the Evaluation of Educational Achievement (IEA).

Nannestad, C. (1998). Concept mapping: The Swiss Army knife of learning, Part 2. *Science and Mathematics Education, 13*, 33-37.

National Centre for Excellence in Teaching Mathematics (NCETM). (2009). *Final report: Researching effective CPD in mathematics education (RECME).* Sheffield: Author. Retrieved from
https://www.ncetm.org.uk/public/files/387088/NCETM+RECME+Final+Report.pdf

National Council of Teachers of Mathematics (NCTM). (1980). *An agenda for action: Recommendations for school mathematics of the 1980s.* Reston, VA: Author.

National Council of Teachers of Mathematics (NCTM). (2000). *Principles and standards for school mathematics.* Reston, VA: Author.

National Council of Teachers of Mathematics (NCTM) Research Committee (2013). New assessments for new standards: The potential transformation of mathematics education and its research implications. *Journal for Research in Mathematics Education,* 44(2), 340-352.

National Mathematics Advisory Panel. (2008). *Foundations for success: The final report of the National Mathematics Advisory Panel.* Washington, DC: Author. Retrieved from http://www.ed.gov/about/bdscomm/list/mathpanel/report/final-report.pdf

National Research Council (NRC). (2000). *How people learn: Brain, mind, experience, and school* (expanded ed.). Committee on the Developments in the Science of Learning. J.D. Bransford, A.L. Brown, & R.R. Cocking, (Eds.). Washington, DC: National Academy Press.

National Research Council (NRC). (2001). *Adding it up: Helping children learn mathematics*. J. Kilpatrick, J. Swafford, & B. Findell, (Eds.). Washington, DC: National Academy Press.

National Research Council & Institute of Medicine. (2004). *Engaging schools: Fostering high school students' motivation to learn.* Washington, DC: National Academy Press.

Nelkon, M., & Parker, P. (1982). *Advanced level physics* (5th ed.). London: Heinemann.

Neyland, A. (1994). Using LOGO in the classroom. In J. Neyland (Ed.), *Mathematics education: A handbook for teachers, Vol. 1* (pp. 65-72). Wellington: Wellington College of Education.

Ng, K.E.D. (2010). Initial experiences of primary school teachers with mathematical modelling. In B. Kaur & J. Dindyal (Eds.), *Mathematical applications and modelling: Yearbook 2010* (pp. 129-147). Singapore: World Scientific.

Ng, K.E.D., & Lee, N.H. (Eds.). (2012). *Mathematical modelling: A collection of classroom tasks*. Singapore: Alston Publishing.

Ng, L.E. (2002). *Representation of problem solving in Singaporean primary mathematics textbooks with respect to types, Polya's model and heuristics.* Unpublished Masters dissertation, Nanyang Technological University, Singapore.

Ng, S.F., & Lee, K. (2009). Model method: A visual tool to support algebra word problem solving at the primary level. In K.Y. Wong, P.Y. Lee, B. Kaur, P.Y. Foong & S.F. Ng (Eds.), *Mathematics education: The Singapore journey* (pp. 169-203). Singapore: World Scientific.

Northrop, E.P. (1944). *Riddles in mathematics: Fun with figures for amateur and expert alike.* Middlesex: Pelican Books.

Noss, R., & Hoyles, C. (1996). *Windows on mathematical meanings: Learning cultures and computers.* Dordrecht: Kluwer Academic.

Novak, J.D. (2010). *Learning, creating, and using knowledge: Concept maps as facilitative tools in schools and corporations* (2nd ed.). New York: Routledge.

Novak, J.D., & Gowin, D. B. (1984). *Learning how to learn.* Cambridge, London: Cambridge University Press.

Ofsted. (2008). *Mathematics: Understanding the score.* London: Author.

Oldknow, A., & Taylor, R. (2000). *Teaching mathematics with ICT.* London: Continuum.

Olivier, A. (1989). Handling students' misconceptions. *Pythagoras, 21*, 10–19.

Ooten, C. (with Moore, K.). (2010). *Managing the mean math blues: Math study skills for student success* (2nd ed.). Upper Saddle River, NJ: Pearson Education.

Oppenheimer, T. (2003). *The flickering mind: The false promise of technology in the classroom and how learning can be saved.* New York, NEW YORK, NY: Random House.

Organisation for Economic Co-operation and Development (OECD). (2005). *Teachers matter: Attracting, developing and retaining effective teachers.* Paris: Author.

Organisation for Economic Co-operation and Development (OECD). (2009). *PISA 2009 assessment framework: Key competencies in reading, mathematics and science.* Paris: Author.

Organisation for Economic Co-operation and Development (OECD). (2013). *PISA 2012 assessment and analytical framework: Mathematics, reading, science, problem solving and financial literacy.* Paris: Author. DOI:10.1787/9789264190511-en

Osgood, C.E., Suci, G.J., & Tannenbaum, P.H. (1957). *The measurement of meaning.* Urbana, IL: University of Illinois Press.

Pape, S. J., & Tchoshanov, M. A. (2001). The role of representation(s) in developing mathematical understanding. *Theory into Practice, 40*(2), 118-127.

Papert, S. (1980). *Mindstorms: Children, computers, and powerful ideas.* New York, NY: Basic Books.

Papert, S. (1993). *The children's machine: Rethinking school in the age of the computer.* New York, NY: Basic Books.

Passmore, T. (2007). Polya's legacy: Fully forgotten or getting a new perspective in theory and practice? *Australian Senior Mathematics Journal, 21*(2), 44-53.

Paulos, J.A. (1995). *A mathematician reads the newspaper.* New York, NY: Basic Books.

Pete, B.M., & Fogarty, R.J. (2003). *Twelve brain principles that make a difference.* Thousand Oaks, CA: Corwin Press.

Piaget, J. (1973). *To understand is to invent: The future of education.* (G.A. Roberts, trans.). New York, NY: Grossman Publishers.

Picker, S.H., & Berry, J.S. (2000). Investigating pupils' images of mathematicians. *Educational Studies in Mathematics, 43,* 65-94.

Pink, D.H. (2009). *Drive: The surprising truth about what motivates us.* New York, NY: Riverhead Books.

Pink, D.H. (2013). *To sell is human: The surprising truth about persuading, convincing, and influencing others.* Edinburgh: Canongate.

Pinker, S. (2014). *The sense of style: The thinking person's guide to writing in the 21st century.* London: Allen Lane.

Poh, B.K. (2007). *Model method: Primary three pupils' ability to use models for representing and solving word problems.* Unpublished Masters dissertation, Nanyang Technological University, Singapore.

Polya, G. (1957). *How to solve it* (2nd ed.). New York, NY: Doubleday & Company.

Polya, G. (ca 1969). *The goals of mathematical education.* Lecture delivered in an inservice course. Retrieved from http://mathematicallysane.com/analysis/polya.asp

Posamentier, A.S., Hartman, H.J., & Kaiser, C. (1998). *Tips for the mathematics teacher: Research–based strategies to help students learn.* Thousand Oaks, CA: Corwin Press.

Posamentier, A.S., & Jaye, D. (2006). *What successful math teachers do, grades 6-12: 79 research-based strategies for the standards-based classroom.* Thousand Oaks, CA: Corwin Press.

Posamentier, A.S., Smith, B.S., & Stepelman, J. (2010). *Teaching secondary school mathematics: Techniques and enrichment units* (8th ed.). Boston: Allyn & Bacon.

Poundstone, W. (2012). *Are you smart enough to work at Google?* New York, NY: Little, Brown and Company.

Prensky, M. (2001). Digital natives, digital immigrants, Part 2: Do they really think differently? *On the Horizon, 9*(6). Retrieved from https://sacs.lamar.edu/prep/QEP/plan/read_res/Prensky_II_Digital_Natives,_Digital_Immigrants_-_Part2.pdf

Prestage, S., & Perks, P. (2001). *Adapting and extending secondary mathematics activities: New tasks for old.* London: David Fulton Publishers.

Public Health England. (2013). *How healthy behaviour supports children's wellbeing.* London: Author. Retrieved from https://www.gov.uk/government/publications/how-healthy-behaviour-supports-childrens-wellbeing

Rajaram, R. (1997). *Differences in performance and choice of heuristics in mathematical problem solving among Secondary Three male gifted pupils in Singapore.* Unpublished Masters dissertation, Nanyang Technological University, Singapore.

Ravitch, D. (2007). *EdSpeak: A glossary of education terms, phrases, buzzwords, and jargon.* Alexandria, VA: Association for Supervision & Curriculum Development.

Rechtschaffen, D. (2014). *The way of mindful education: Cultivating well-being in teachers and students.* New York, NY: W. W. Norton & Co.

Robitaille, D.F., & Garden, R.A. (Eds.). (1996). *Research questions & study design. TIMSS Monograph No. 2.* Vancouver: Pacific Educational Press.

Rohrer, D. (2009). The effects of spacing and mixing practice problems. *Journal for Research in Mathematics Education, 40*(1), 4-17.

Romagnano, L. (2006). *Mathematics assessment literacy: Concepts and terms in large-scale assessment.* Reston, VA: National Council of Teachers of Mathematics.

Rosenshine, B. (2012). Principles of instruction: Research-based strategies that all teachers should know. *America Educator, 36*(1), 12-39.

Rothstein, D., & Santana, L. (2014). The right questions. *Educational Leadership online, 72*(2). Retrieved from http://www.ascd.org/publications/educationalleadership/oct14/vol72/num02/TheRightQuestions.aspx

Rowe, M.B. (1978). *Teaching science as continuous enquiry.* New York, NY: McGraw-Hill.

Rowe, M. (1987). Wait-time: Slowing down may be a way of speeding up. *American Educator, 11*, 38-43.

Royal Society & Joint Mathematical Council (2001). *Teaching and learning geometry 11-19: Report of a Royal Society/JMC working group chaired by A. Oldknow.* London: The Royal Society.

Ryan, J., & Williams, J. (2007). *Children's mathematics 4 – 15.* Berkshire: McGraw-Hill and Open University Press.

Ryan, J., & Williams, J. (2010). Children's mathematical understanding as a work in progress: Learning from errors and misconceptions. In I. Thompson (Ed.), *Issues in teaching numeracy in primary schools* (2nd ed.) (pp. 146-157). Buckingham: Open University Press.

Sarason, S.B. (1993). *You are thinking of teaching? Opportunities, problems, realities.* San Francisco, CA: Jossey-Bass Publishers.

Schaffhauser, D. (2014). Hybrid classes outlearn traditional classes. *THE Journal.* Retrieved from http://thejournal.com/articles/2014/12/18/hybrid-classes-outlearn-traditional-classes.aspx

Schoenfeld, A.H. (1987). What's all the fuss about metacognition? In A.H. Schoenfeld (Ed.), *Cognitive science and mathematics education* (pp. 189-215). Hillsdale, NJ: Lawrence Erlbaum Associates.

Schoenfeld, A.H. (Ed.). (2008). *A study of teaching: Multiple lenses, multiple views.* Reston, VA: National Council of Teachers of Mathematics.

Schoenfeld, A.H. (2010). Reflections of an accidental theorist. *Journal for Research in Mathematics Education, 41*(2), 104-116.

Scholtz, A. (2011). The regions in a circle (RC) formula. *Mathematics Teacher, 105(3),* 167-168.

Shannon, A. (2007). Task context and assessment. In Alan H. Schoenfeld (Ed.), *Assessing mathematical proficiency* (pp. 177-191). New York, NY: Cambridge University Press.

Shulman, L.S. (1986). Those who understand: Knowledge growth in teaching. *Educational Researcher, 15*(2), 4-14.

Siraj-Blatchford, I., Shepherd, D., Melhuish, E., Taggart, B., Sammons, P., & Sylva, K. (2011). *Effective Primary Pedagogical Strategies in English and Mathematics in Key Stage 2: A study of Year 5 classroom practice from the EPPSE 3-16 longitudinal study.* London: Institute of Education, Birkbeck (University of London) and University of Oxford.

Skemp, R. (1971). *The psychology of learning mathematics.* Middlesex, UK: Penguin Books.

Skemp, R.R. (1976). Relational understanding and instrumental understanding. *Mathematics Teaching, 77*, 20-26.

Skemp, R. (1979). Goals of learning and qualities of understanding. *Mathematics Teaching, 88*, 44-49.

Sousa, D.A. (2008). *How the brain learns mathematics*. Thousand Oaks, CA: Corwin Press.

Stahl, R.J. (1994). *Using "think-time" and "wait-time" skilfully in the classroom: ERIC Digest.* Retrieved from ERIC database. (ED370885)

Stewart, I. (2006). *Letters to a young mathematician.* New York, NY: Basic Books.

Stewart, I. (2009). *Professor Steward's cabinet of mathematical curiosities.* New York, NY: Basic Books.

Stewart, I. (2011). *The mathematics of life.* New York, NY: Basic Books.

Stillman, G.A., Kaiser, G, Blum, W., & Brown, J.P. (Eds.). (2013). *Teaching mathematical modelling: Connecting to research and practice.* Dordrecht, Netherlands: Springer. doi:10.1007/978-94-007-6540-5_19

Sweller, J. (1992). Cognitive theories and their implications for mathematics instruction. In G. Leder (Ed.), *Assessment and learning of mathematics* (pp. 46-62). Hawthorn, Victoria: Australian Council for Educational Research.

Sweller, J., van Merrienboer, J.J.G., & Paas, F.G.W.C. (1998). Cognitive architecture and instructional design. *Educational Psychology Review, 10*(3), 251-296.

Swetz, F., & Hartzler, J.S. (Eds.). (1991). *Mathematical modeling in the secondary school curriculum: A resource guide of classroom exercises.* Reston, VA: National Council for Teachers of Mathematics.

Tall, D. (1989). Concept images, generic organizers, computers, and curriculum change. *For the Learning of Mathematics, 9*(3), 37-42.

Tall, D. & Vinner, S. (1981). Concept image and concept definition in mathematics with particular reference to limits and continuity. *Educational Studies in Mathematics,* 12, 151-169.

Tan, K.S.S., & Goh, C.B. (Eds.). (2003). *Securing our future: Sourcebook for infusing National Education into the primary school curriculum.* Singapore: Pearson Education Asia.

Tapper, J. (2012). *Solving for why: Understanding, assessing, and teaching students who struggle with math, Grades K-8.* Sausalito, CA: Math Solutions.

Tatto, M.T., Schwille, J., Senk, S.L., Ingvarson, L., Rowley, G., Peck, R., Bankov, K., Rodriguez, M., & Reckase, M. (2012). *Policy, practice, and readiness to teach primary and secondary mathematics in 17 countries: Findings from the IEA Teacher Education and Development Study in Mathematics (TEDS-M).* Amsterdam, the Netherlands: International Association for the Evaluation of Educational Achievement (IEA).

Tay, E.G., Quek, K.S., & Toh, T.L. (2011). Affective assessment in the mathematics classroom: A quick start. In B. Kaur & K.Y. Wong (Eds.), *Assessment in the mathematics classroom: Association of Mathematics Educators Yearbook 2011* (pp. 257-273). Singapore: World Scientific.

Taylor, R.P. (1980). Introduction. In R.P. Taylor (Ed.), *The computer in school: Tutor, tool, tutee* (pp. 1-10). New York, NY: Teachers College Press. Retrieved from http://www.citejournal.org/vol3/iss2/seminal/article1.cfm

Teng, A. (2014, November 24). Ex-teacher Kho Tek Hong solved Singapore's maths problem: Ex-teacher led team that created model method for teaching, learning. *The Straits Times,* p.B7.

Teong, S.K., Hedberg, J.G., Ho, K.F., Lioe, L.T., Tiong, J., Wong, K.Y., & Fang, Y.P. (2009). *Developing the repertoire of heuristics for mathematical problem solving: Project 1* (Unpublished Technical Report). Singapore: National Institute of Education, Centre for Research in Pedagogy and Practice.

Thomas, M.O.J. (2008). Developing versatility in mathematical thinking. *Mediterranean Journal for Research in Mathematics Education, 7*(2), 67-87.

Toh, T.L., Quek, K.S., Leong, Y.H., Dindyal, J., & Tay, E.G. (2011). *Making mathematics practical: An approach to problem solving.* Singapore: World Scientific.

Toh, T.L., Quek, K.S., & Tay, E.G. (2008). *Problem solving in the mathematics classroom (Junior College).* Singapore: National Institute of Education and Association of Mathematics Educators.

Tough, P. (2013). *How children succeed: Grit, curiosity, and the hidden power of character.* Boston, MA: Mariner Book.

Tufte, E.R. (2003). *The cognitive style of PowerPoint.* Cheshire, CT: Graphics Press.

Usiskin, Z. (2001). Teachers' mathematics: A collection of content deserving to be a field. *The Mathematics Educator, 6*(1), 86-98.

Usiskin, Z., & Griffin, J. (with Willmore, E. & Witonsky, D.). (2008). *The classification of quadrilaterals: A study of definition.* Charlotte, NC: Information Age Publishing.

Valverde, G.A., Bianchi, L.J., Wolfe, R.G., Schmidt, W.H., & Houang, R.T. (2002). *According to the book: Using TIMSS to investigate the translation of policy into practice through the world of textbooks.* Dordrecht: Kluwer Academic Publishers.

van Hiele, P.M. (1986). *Structure and insight: A theory of mathematics education.* Orlando, FL: Academic Press.

Verschaffel, L., Greer, B., & De Corte, E. (2000). *Making sense of word problems.* Lisse, the Netherlands: Swets & Zeitlinger Publishers.

Walker, L. (2011). *Model drawing for challenging word problems: Finding solutions the Singapore way.* Peterborough, NH: Crystal Springs Books.

Walsh, J.A., & Sattes, B.D. (2005). *Quality questioning: Research-based practice to engage every learner.* Thousand Oaks, CA: Corwin Press.

Walters, K., Smith, T.M., Leinwand, S., Surr, W., Stein, A., & Bailey, P. (2014). *An up-close look at student-centered math teaching: A study of highly regarded high school teachers and their students.* Washington, DC: American Institutes for Research.

Waring, S. (2000). *Can you prove it? Developing concepts of proof in primary and secondary schools.* Leicester: Mathematical Association.

Waters, H.S., & Schneider, W. (Eds.). (2010). *Metacognition, strategy use, & instruction.* New York, NY: Guilford Press.

Watkins, D. (2007). Comparing ways of learning. In M. Bray, B. Adamson & M. Mason (Eds.), *Comparative education research: Approaches and methods* (pp. 299-313). Hong Kong: University of Hong Kong.

Watson, A., & Mason, J. (2005). *Mathematics as a constructive activity: Learners generating examples.* Mahwah, NJ: Lawrence Erlbaum Associates.

Wenglinsky, H. (2002). How schools matter: The link between teacher classroom practices and student academic performance. *Education Policy Analysis Archives, 10*(12). Retrieved from http://epaa.asu.edu/epaa/v10n12/

Wiggins, G.P., & McTighe, J. (2011). *The Understanding by Design guide to creating high quality units.* Alexandria, VA: ASCD.

Wiliam, D. (2011). *Embedded formative assessment.* Bloomington, IN: Solution Tree Press.

Wilkins, J.M. (2004). Mathematics and science self-concept: An international investigation. *Journal of Experimental Education, 72*(4), 331-346.

Willis, J. (2007). The neuroscience of joyful education. *Educational Leadership, 64* (online only).

Winter, J. (2001). Personal, spiritual, moral, social and cultural issues in teaching mathematics. In P. Gates (Ed.), *Issues in mathematics teaching* (pp. 197-213). London: RoutledgeFalmer.

Wong, K.Y. (1979). A statistical study of Bahasa Malaysia. *Berita Matematik, 17*, 11-19.

Wong, K.Y. (1984). *Mathematical understanding: An exploration of theory and practice.* Unpublished doctoral thesis, University of Queensland, Australia.

Wong, K.Y. (1987). Aspects of Mathematical Understanding. *Singapore Journal of Education, 8*(2), 45-55.

Wong, K.Y. (1990a). A graphing program using Logo. *Journal of Computers in Mathematics and Science Teaching, 9*(2), 85-91.

Wong, K.Y. (1990b, July). *Mathematical vitality revisited: A comparative study.* Paper presented at MERGA (Mathematics Education Research Group of Australasia) Annual Conference, Hobart.

Wong, K.Y. (1991). Strategy approach to learning mathematics. In K.V. Palanisamy, F. Lopez-Real, & S. Terlochan (Eds.), *Proceedings of the Fifth South East Asian Conference on Mathematical Education* (pp. 22-30). Brunei: Universiti Brunei Darussalam.

Wong, K.Y. (1996). Assessing perceptions using student drawings. In M. Quigley, P.K. Veloo, & K.Y. Wong (Eds.), *Assessment and evaluation in science and mathematics education: Innovative approaches* (pp. 370-379). Brunei Darussalam: Universiti Brunei Darussalam.

Wong, K.Y. (1997a). 3.14159 26535 89793 23846 26433 83279 50288 41971 ... *Science and Mathematics Education, 11*, 17-22.

Wong, K.Y. (1997b). *Computers for Mathematics Instruction (CMI) Project, Module 1: Excel mathematics: Using templates.* Brunei Darussalam: Universiti Brunei Darussalam.

Wong. K.Y. (1997c). Student teachers' construction of content knowledge in secondary mathematics. In L.C. Arañador, I.N. Valencia, & T. Vui (Eds.), *Proceedings of 1997*

International Conference on Cooperative Learning and Constructivism in Science and Mathematics Education (2-1 to 2-6). Penang: SEAMEO RECSAM.

Wong, K.Y. (1998a). *Computers for Mathematics Instruction (CMI) Project, Module 2: Graphing software.* Brunei Darussalam: Universiti Brunei Darussalam. Available from http://academic.sun.ac.za/mathed/MALATI/Graphmat.pdf

Wong, K.Y. (1998b). Thinking hats and mathematics homework. *Science and Mathematics Education, 13,* 38-44.

Wong, K.Y. (1999). Multi-modal approach of teaching mathematics in a technological age. In E. B. Ogena & E. F. Golia (Eds.), *8th Southeast Asian Conference on Mathematics Education, technical papers: Mathematics for the 21st century* (pp. 353-365). Manila: Ateneo de Manila University.

Wong, K.Y. (2000a). *Computers for Mathematics Instruction (CMI) Project, Module 3: Dynamic geometry.* Brunei Darussalam: Universiti Brunei Darussalam.

Wong, K.Y. (2000b). In-situ reflections on fractals. In K.A. Clements, Hassan H. Tairab, & K.Y. Wong (Eds.), *Science, mathematics and technical education in the 20th and 21st centuries* (pp. 236-245). Bandar Seri Begawan: Department of Science and Mathematics Education, SHBIE, Universiti Brunei Darussalam.

Wong, K.Y. (2001a). *Computers for Mathematics Instruction (CMI) Project, Module 4: Logo mathematics.* Brunei Darussalam: Universiti Brunei Darussalam.

Wong, K.Y. (2001b). Nine propositions about future-oriented mathematics education (FOME): An introduction. *The Mathematics Educator, 6*(1), 22-41.

Wong, K.Y. (2002a). Helping your students to become metacognitive in mathematics: A decade later. *Mathematics Newsletter, 12*(5). Available from http://math.nie.edu.sg/kywong/Metacognition%20Wong.pdf

Wong, K.Y. (2002b). Mathematics for science: A Brunei experience. *Teaching Mathematics and Its Applications, 21*(2), 55-65.

Wong, K.Y. (2003a). Enhancing students' learning through error analysis. *Science, Mathematics and Technical Education, 18,* 41-46.

Wong, K.Y. (2003b). Mathematics-based national education: A framework for instruction. In K.S.S. Tan & C.B. Goh (Eds.), *Securing our future: Sourcebook for infusing National Education into the primary school curriculum* (pp. 117-130). Singapore: Pearson Education Asia.

Wong, K.Y. (2004). Reviving the roles of Logo in the teaching of mathematics. In A.H. Yahya, B. Adam, A.I. Ismail, H.L. Khoh & H.C. Low (Eds.), *Integrating technology in the mathematical sciences* (pp. 18-27). Penang: Universiti Sains Malaysia.

Wong, K.Y. (2005, August). *A conceptualization of pre-service mathematics teacher education: Framework and principles of teacher training.* Paper presented at the Third East Asia Regional Conference on Mathematics Education (ICMI-EARCOME 3), Shanghai, China.

Wong, K.Y. (2008a). An extended Singapore mathematics curriculum framework. *Maths Buzz, 9*(1), 2-3.

Wong, K.Y. (2008b). Success and consistency in the use of heuristics to solve mathematics problems. In M. Goos, R. Brown, & K. Makar (Eds.), *Navigating currents and charting directions* (Proceedings of the 31st Annual Conference of the Mathematics Education Research Group of Australasia) Vol. 2 (pp. 589-595). Adelaide: MERGA.

Wong, K.Y. (2009a). ICT and mathematics education. In P.Y. Lee & N.H. Lee (Eds.), *Teaching secondary school mathematics: A resource book (2nd. & updated ed.)* (pp. 357-368). Singapore: McGraw-Hill Education (Asia).

Wong, K.Y. (2009b). Introduction: A framework for learning to teach mathematics in primary schools. P.Y. Lee & N.H. Lee (Eds.), *Teaching primary school mathematics: A resource book (2nd. ed.)* (pp. 3-14). Singapore: McGraw-Hill Education (Asia).

Wong, K.Y. (2012). Use of student mathematics questioning to promote active learning and metacognition. In *Electronic compilation of the 12th International Congress on Mathematical Education* (ICME-12) (pp. 1086-1100). Seoul, Korea.

Wong, K.Y. (2013a, January). *Diverse pathways for life-long teacher professional development.* Special, invited presentation delivered at International Science, Mathematics and Technology Education Conference (ISMTEC 2013), Bangkok, Thailand.

Wong, K.Y. (2013b). Metacognitive reflection at secondary level. In B. Kaur (Ed.), *Nurturing reflective learners in mathematics: Association of Mathematics Educators Yearbook 2013* (pp. 81-102). Singapore: World Scientific.

Wong, K.Y. (2014). M_Crest: A framework of motivation to learn mathematics. In P.C. Toh, T.L. Toh, & B. Kaur (Eds.), *Learning experiences to promote mathematics learning: Association of Mathematics Educators Yearbook 2014* (pp. 13-40). Singapore: World Scientific.

Wong, K.Y., Boey, K.L., Lim-Teo, S.K., & Dindyal, J. (2012). The Preparation of Primary Mathematics Teachers in Singapore: Programs and Outcomes from the TEDS-M study. *ZDM: The International Journal on Mathematics Education, 44*(3), 293-306. doi: 10.1007/s11858-011-0370-1

Wong, K.Y., & Chen, Q. (2012). Nature of an Attitudes toward Learning Mathematics Questionnaire. In J. Dindyal, L.P. Cheng & S.F. Ng (Eds.), *Mathematics education: Expanding horizons: Proceedings of the 35th annual conference of the Mathematics Education Research Group of Australasia* (pp. 793-800). Adelaide: MERGA.

Wong, K.Y., Kaur, B., Koay, P.L., & Jamilah Mohd Yusof (2009). My "best" mathematics teacher: Perceptions of primary school pupils from Singapore and Brunei Darussalam. In K.Y. Wong, P.Y. Lee, B. Kaur, P.Y. Foong & S.F. Ng (Eds.), *Mathematics education: The Singapore journey* (pp. 512-524). Singapore: World Scientific.

Wong, K.Y., & Lee, N.H. (2009). Singapore education and mathematics curriculum. In K.Y. Wong, P.Y. Lee, B. Kaur, P.Y. Foong & S.F. Ng (Eds.), *Mathematics education: The Singapore journey* (pp. 13-47). Singapore: World Scientific.

Wong K.Y., Lim, Y.S., & Loh, K.G. (1989). *Computers in Education Project Report: Country report: Singapore*. Penang, Malaysia: Regional Centre for Education in Science and Mathematics (RECSAM).

Wong, K.Y., & Low-Ee, H.W. (2006). In-class experiences of teachers and students during mathematics lessons: Case studies at Primary and Polytechnic levels. In W.D. Bokhorst-Heng, M.D. Osborne, & K. Lee (Eds.), *Redesigning pedagogy: Reflections on theory and praxis* (pp. 101-117). Rotterdam: Sense Publishers.

Wong, K.Y., Oh, K.S., Ng, Q.T., & Cheong, S.K. (2012, July). *Use of internet in a mathematics assessment system with semi-automatic marking and customisable feedback*. Paper presented at Topic Study Group 18 of the 12th International Congress on Mathematical Education (ICME-12). Seoul, Korea. Available from http://hdl.handle.net/10497/6208

Wong, K.Y., & Quek, K. S. (2009). *Enhancing Mathematics Performance (EMP) of mathematically weak pupils: An exploratory study* (Unpublished Technical Report). Singapore: National Institute of Education, Centre for Research in Pedagogy and Practice.

Wong, K.Y., & Quek, K.S. (2010). Promote student questioning in mathematics lessons. *Maths Buzz, 11*(1), 2-3.

Wong, K.Y., & Tiong, J. (2008). *Developing the Repertoire of Heuristics for Mathematical Problem Solving: Student Problem Solving Exercises and Attitude* (Unpublished Technical Report). Singapore: National Institute of Education, Centre for Research in Pedagogy and Practice.

Wong, K.Y., & Veloo, P.K. (1996). Multimodal teaching strategy to promote understanding of secondary mathematics. In H. Forgasz, T. Jones, G. Leder, J. Lynch, K. Maguire & C. Pearn (Eds.), *Mathematics: Making connections* (pp. 258-264). Melbourne: Mathematical Association of Victoria.

Wong, K.Y., & Veloo, P. (1997). A typology of pupils' types by beliefs about mathematics learning. In L.K. Chen & K.A. Toh (Eds.), *Educational Research Association 1997 Annual Conference Proceedings: Research across the disciplines* (pp. 307-316). Singapore: Educational Research Association.

Wong, K.Y., Zaitun Mohd Taha, & Veloo, P. (2001). Situated sociocultural mathematics education: Vignettes from Southeast Asian practices. In B. Atweh, H. Forgasz & B. Nebres (Eds.), *Sociocultural research on mathematics education: An international research perspective* (pp. 113-134). Mahwah, NJ: Lawrence Erlbaum.

Wong, K.Y., Zhao, D.S., Cheang, W.K., Teo, K.M., Lee, P.Y., Yen, Y.P., Fan, L.H., Teo, B.C., Quek, K.S., & So, H.J. (2012). *Real-life mathematics tasks: A Singapore experience*. Singapore: Centre for Research in Pedagogy and Practice, National Institute of Education, Nanyang Technological University.

Wong, S.O. (2002). *Effects of heuristics instruction on pupils' achievement in solving non-routine problems*. Unpublished Masters dissertation, Nanyang Technological University, Singapore.

Wong, S.O., & Lim-Teo, S.K. (2002). Effects of heuristics instruction on pupils' mathematical problem-solving process. In D. Edge & B.H. Yeap (Eds.), *Proceedings of Second East Asia Regional Conference on Mathematics Education & Ninth Southeast Asian Conference on Mathematics Education* (pp. 180-186). Singapore: Association of Mathematics Educators.

Wood, T. (1998). Alternative patterns of communication in mathematics classes: Funneling or focusing? In H. Steinbring, M. G. Bartolini Bussi & A. Sierpinska (Eds.), *Language and communication in the mathematics classroom* (pp. 167-178). Reston, VA: National Council of Teachers of Mathematics.

Woollard, J. (2011). *Psychology for the classroom: E-learning.* Oxon: Routledge.

World Bank. (2015). *World Development Report 2015: Mind, Society, and Behavior.* Washington, DC: Author. doi: 10.1596/978-1-4648-0342-0

Wragg, E.C., & Brown, G. (2001). *Questioning in the primary school.* London: Routledge/Falmer.

Wu, H.H. (2011). *Understanding numbers in elementary school mathematics.* Providence, RI: American Mathematical Society.

Wu, Y.K. (2005). *Statistical graphs: Understanding and attitude of Singapore secondary school students and the impact of a spreadsheet exploration.* Unpublished PhD thesis, Nanyang Technological University, Singapore.

Xin, Y.P. (2012). *Conceptual model-based problem solving: Teach students with learning difficulties to solve math problems.* Rotterdam: Sense Publishers.

Yan, K.C. (2002). The model method in Singapore. *The Mathematics Educator, 6*(2), 47-62.

Yeap, B.H. (2008). *Problem solving in the mathematics classroom (Primary).* Singapore: National Institute of Education and Association of Mathematics Educators.

Yeap, B.H. (2010). *Bar modelling: A problem solving tool. From research to practice.* Singapore: Marshall Cavendish Education.

Yeo, J.B.W. (2010). Why study mathematics? Applications of mathematics in our daily life. In B. Kaur & J. Dindyal (Eds.), *Mathematical applications and modelling: Yearbook 2010* (pp. 151-177). Singapore: World Scientific.

Zazkis, R. (2011). *Relearning mathematics: A challenge for prospective elementary school teachers.* Charlotte, NC: Information Age Publishing.

Zenner, C, Herrnleben-Kurz, S., & Walach, H. (2014). Mindfulness-based interventions in schools: A systematic review and meta-analysis. *Frontiers in Psychology, 5*(603). Retrieved from http://journal.frontiersin.org/Journal/10.3389/fpsyg.2014.00603/full#

Zhao, D.S., Cheang, W.K., Teo, K.M., & Lee, P.Y. (2011). Some principles and guidelines for designing mathematical disciplinary tasks for Singapore schools. In J. Clark, B. Kissane, J. Mousley, T. Spencer & S. Thornton (Eds.), *Mathematics: Traditions and (new) practices: Proceedings of the AAMT-MERGA conference* (pp. 1107-1115). Adelaide: Australian Association of Mathematics Teachers and Mathematics Education Research Group of Australasia.

Printed in the United States
By Bookmasters

Printed in the United States
By Bookmasters